Statistical Modelling

Statistical Modelling

Warren Gilchrist

Head of the Department of Mathematics,
Statistics, and Operational Research,
Sheffield City Polytechnic, UK

JOHN WILEY & SONS

Chichester · New York · Brisbane · Toronto · Singapore

Library of Congress Cataloging in Publication Data:

Gilchrist, Warren, 1932–
  Statistical modelling.

  Includes index.
  1. Linear models (Statistics) I. Title.
QA276.G453 1984   519.5   83–21584
ISBN 0 471 90380 9 (U.S.)
ISBN 0 471 90391 4 (U.S. : pbk.)

British Library Cataloguing in Publication Data:

Gilchrist, Warren
  Statistical modelling.
  1. Statistics
  I. Title
  519.5   QA276

  ISBN 0 471 90380 9
  ISBN 0 471 90391 4 pbk

Phototypeset by Macmillan India Ltd and printed in Great Britain by Page Bros (Norwich) Ltd, Norwich

To my Father

Contents

PART III IDENTIFICATION

PART V VALIDATION

PART VI APPLICATION

PART VII ITERATION

Preface

The construction and use of quantitative models in many areas of the social sciences has been a major development over the last thirty to forty years. Their application to science and engineering go back much further, though even here there have been great strides in recent years. Much of the emphasis has been on the mathematics of these models. However, a purely mathematical approach runs the danger of developing models that are unrealistic. To relate models to reality requires a consideration of the evidence presented by observation and an allowance for the ever-present uncertainty in the world. The study of the roles of data and of uncertainty in modelling is the subject matter of statistical modelling. The aim of this introductory book is to bring together and present in a structured fashion those statistical and related ideas that are needed for the construction of quantitative models. The structure of the book is based on consideration of the processes involved in modelling and the role of statistics within each of these. This has led to a structure based on seven parts:

Part I	The Modelling Process
Part II	Common Statistical Models
Part III	Identification
Part IV	Estimation
Part V	Validation
Part VI	Application
Part VII	Iteration

A foundation of statistical ideas and techniques is laid in Parts I and II, then the statistical methods needed within the component processes of modelling are examined in Parts III to VII.

The intention has been to make this book self-contained. However, as the focus is on modelling, some statistical topics, such as probability and significance tests, are only introduced briefly before being applied to modelling. The reader who is totally unfamiliar with statistics might find it helpful to seek further illustration from other statistical texts.

To carry out modelling at an advanced level often requires advanced mathematical techniques. However, the concepts underlying these techniques are often based on much simpler principles. These can be explained and illustrated by reference to simple problems. The objective of this book is to emphasize the underlying concepts and to illustrate them by simple examples that make these concepts clear. The level of mathematics required has therefore been limited to basic algebra with rare references to differentiation. The reader should be aware that to apply the concepts to more sophisticated examples will require more advanced mathematical techniques. Where topics of importance involve mathematics or theory beyond the scope of this introductory text some general comments have been made and appropriate references given at the end of the book, so the interested reader may follow up the topics elsewhere. The References sections therefore contain some suggested reading as well as detailing references given in the text. The chapters and sections involving relatively difficult ideas are indicated by a * and might be omitted for initial study.

The author has had difficulty with the choice of words to describe various aspects of the subject. The literature is full of different words applied to the same idea and also the same words applied to totally different ideas. The greatest difficulty has been experienced over the three terms that appear frequently in chapter headings. They are the phrases: *the Conceptual Approach*, which is the theory-based approach to modelling, called in other texts by terms such as 'synthesis'; *the Empirical Approach*, which is the approach based on the analysis of data; and *the Eclectic Approach*, which is the approach that seeks to weld together conceptual and empirical elements. The word 'eclectic' is often used in a rather derogatory fashion, but it means 'selecting from various alternatives those parts that are useful and good'. This is precisely what one is trying to do in modelling and therefore, though the author has endeavoured to find an alternative, he has decided to use this word as the most appropriate.

Acknowledgements

The author would like to express his appreciation to the following people who have commented on all or parts of the draft of the book: J. Bryant, D. R. Cox, D. Drew, R. Harris, C. Mallows, G. Morgan. He would like to thank Mrs C. Barker and Ms S. Simms for their careful typing. Above all he would like to thank his father, who, at past eighty, has still found the energy and enthusiasm to try to get his son to write more clearly. That he has inevitably failed in places is the author's fault and not his.

Notation

Capitals	X, Y, R etc.	—Random variables
Small	x, y, r, etc.	—Observed values of random variables
Greek letters	$\alpha, \beta, \mu, \sigma$, etc.	—Parameter values
	$\hat{\alpha}, \hat{\beta}, \hat{\mu}, \hat{\sigma}$, etc.	—Estimates of parameters

$\ln x$ Natural logarithms, log to the base e

$\cap$ is distributed as, has the distribution $\sqrt{}$

$\simeq$ Approximately equal to

$\propto$ Proportional to

$\in$ Belongs to

$\left(\dfrac{dy}{dx}\right)_{x_0}$ A rate of change, $\dfrac{dy}{dx}$, calculated at $x = x_0$

$\displaystyle\sum_i$ Summation over all values taken by i

$\displaystyle\prod_i$ Product over all values taken by i

$\dfrac{\partial y}{\partial x}, \dfrac{\partial y}{\partial z}$ Rates of chage of y with respect to x and z in each case holding the other constant

$E(.)$ Expectation of

$V(.)$ Variance of

Part I

The modelling process

Chapter 1
An overview

1.1 An introductory example

A range of problems related to the positioning of stores and the planning of delivery routes requires information on the distances by road, y, between different places. Where a large number of such places are involved, finding these distances by driving or by direct measurement along the roads on a map is time-consuming. To avoid this problem the usual approach is to relate the road distances to the straight line distance, denoted by x, as measured using a scale map. This relationship will be expressed mathematically and will enable us to predict a value of y given a corresponding value of x. This relationship will be our quantitative model of the situation. The fundamental question is: how do we obtain this relationship. There exists a whole spectrum of differing approaches. It is helpful in discussing these approaches to identify the two extremes of this spectrum and afterwards study the remainder of the spectrum which usually consists of a combination of the extremes. Let us do this for our example. One extreme approach to modelling is to derive the form of the model on the basis of our understanding of the situation. We thus use logical reasoning, 'known theory', to obtain the model. We shall refer to this approach as the *Conceptual* approach and consider its application to the problem in hand. Considering the situation we can write the following statements:

(a) When $x = 0$ the two points coincide, so $y = 0$.
(b) If there is a straight road between the two points then $y = x$, but otherwise y will be greater than x. Mathematically we write $y \geqslant x$.
(c) The distance y will generally increase with x. However, the randomness in the pattern of roads will mean that even if different pairs of places have the same x there will be random differences in the y values. Thus each y will consist of a part predictable from the x value and an unpredictable, random, component, which we shall denote by e.
(d) Provided we keep to relatively similar situations, e.g. urban roads, then

3

the form of the relationship should not depend strongly on the distances involved. Thus if the straight line distances are doubled we would expect in most cases the road distances also to be approximately doubled.

We now seek the simplest relationship that satisfies these conditions. Consider a few possibilities:

(i) $y = x$: satisfies (a) and (d) but not (b) or (c).
(ii) $y = x +$ random component: this now allows (c) but not (b).
(iii) $y = x +$ constant + random component: this helps with (b) but (a) now fails.
(iv) $y =$ constant . $x +$ random component: this now satisfies all four provided that the constant is greater than one and that the random component, e, is constrained so that $y \geqslant x$. Keeping to the convention that constants, parameters, are symbolized by Greek symbols we shall write our 'conceptual' model as

$$y = \theta x + e, \qquad \theta \geqslant 1, \quad y \geqslant x.$$

Notice that this model is derived without any data being provided about the actual situation.

The other extreme approach is to consider only empirical evidence and to ignore the evidence we have from previous experience or knowledge. This approach we shall term the *Empirical* approach to modelling. To apply this approach to the problem we begin with data. Table 1.1. gives twenty observations on road and straight line distances for twenty pairs of points in Sheffield. These points are plotted in Figure 1.1. An examination of this figure suggests that a straight line through the origin follows the main trend of the data, though clearly there is also a random component. Thus if the observation (x_i, y_i) represents one of the points the relation can be expressed as

$$y_i = \hat{\theta} x_i + \hat{e}_i, \qquad i = 1, 2, \ldots, 20,$$

where $\hat{\theta}$ is the slope of the drawn line and $\hat{e}_i$ is the error in predicting y_i from the line. The form of this model is identical to that obtained by conceptual argument. Without the subscripts the model is taken as appropriate for distances x other than those of the specific observations:

$$y = \theta x + e.$$

We have now arrived at the same model by both the conceptual and empirical approaches. The conceptual approach used what we might term *Prior Information*, i.e. the information that we possessed prior to obtaining the data. The empirical approach ignored this information and used only the *Empirical Information* contained in the data. Clearly in practice we should seek to use both sources of information. In our example, the fact

Table 1.1 Road and straight line distances in Sheffield

(a) Road distance	Linear distance	Road distance	Linear distance
y	x	y	x
10.7	9.5	11.7	9.8
6.5	5.0	25.6	19.0
29.4	23.0	16.3	14.6
17.2	15.2	9.5	8.3
18.4	11.4	28.8	21.6
19.7	11.8	31.2	26.5
16.6	12.1	6.5	4.8
29.0	22.0	25.7	21.7
40.5	28.2	26.5	18.0
14.2	12.1	33.1	28.0

(b) Linear distance	Residual	Linear distance	Residual
x	$y-1.3x$	x	$y-1.3x$
4.8	-0.26	15.2	2.56
5.0	6.00	18.0	-3.10
8.3	1.29	19.0	-0.90
9.5	1.65	21.6	-0.72
9.8	1.04	21.7	2.51
11.4	-3.58	22.0	-0.40
11.8	-4.36	23.0	0.50
12.1	-0.87	26.5	3.25
12.1	1.53	28.0	3.30
14.6	2.68	28.2	3.84

(c) *Frequency table*

Interval	Frequency
$e < -5$	0
$-5 \leqslant e < -3$	4
$-3 \leqslant e < -1$	0
$-1 \leqslant e < 1$	7
$1 \leqslant e < 3$	7
$3 \leqslant e < 5$	2
$5 \leqslant e$	0

that both sources of information led to the same form of model adds greatly to our confidence in that model. Even had they not done so we might have been able to use the prior information to guide us in checking the empirical information and vice versa. This combined approach we shall term the *Eclectic* approach.

Having obtained the model, we shall wish to put it to use to predict road distances from straight line distances in Sheffield. Here the conceptual approach tells us only that $\theta > 1$. We might have a report that tells us that

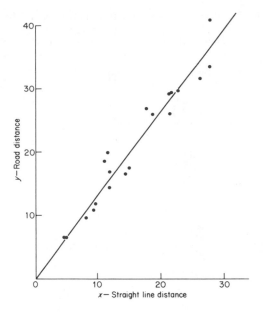

Figure 1.1 Plot of road and straight line distances

for some other city $\theta = 1.5$; however, there is no evidence that different cities have the same θ. At this stage therefore we have to take an empirical approach and use our data to provide a numerical value for θ. This value we refer to as an estimate of θ. We later obtain some methods and formulae for obtaining this, but a line drawn by eye will in this case give us a fair idea. The line fitted by eye is approximately

$$y = 1.3\,x.$$

If we measure the mileages by vehicles between points, over a number of trials we could see how well the model predicts these mileages. This will enable us to test the validity of the model.

The model as it has been defined concentrates on the line and provides no information about the random component, e. As a further stage in model construction we must consider how to model the properties of this random component. We shall develop the techniques of doing this in later chapters. For the moment it is sufficient to examine the evidence of the twenty observations. The errors from the fitted line will be $y_i - 1.3\,x_i$. Table 1.1(b) gives these errors, called *Residuals*. Table 1.1(c) gives a summary 'frequency' table, which shows the sort of raggedness that one must expect with only twenty observations. Nonetheless we can make some general observations. The errors tend to balance between positive and negative values and there is no obvious increase in the magnitude of the errors for increasing distance. If we wished to use the model for prediction purposes, for x in the range 5 to 30, then we would expect to have most errors within ± 4 units. We thus can see that to discuss the random component of the model a different language is needed from the mathematical statements of the deterministic component. This is the language of expectations and probabilities, the language of statistics.

1.2 Models

If we think how, when growing during childhood, we learned to understand the world we soon realise the role of models. Model trains, model aeroplanes, dolls, building sets, were central to our learning. Models of situations, simulations, appeared as games of playing at doctors or nurses, teachers, cops and robbers. We could not deal with the real thing so a model had to suffice. The real thing was too inaccessible, too big, too complicated, too expensive, too dangerous. As adults, for much the same reasons, we build models of river estuaries to study the effects of a barrage, of aeroplanes to fly in wind tunnels; similarly we play business games, war games, and peace games. Many of the models we have referred to so far involve physical things. The application of models can, however, be greatly extended if we build models using symbols instead of things. Models of national economies have been built using water to represent the flows of wealth. It is easier and more flexible to use different mathematical symbols to represent the multitude of components of the economy, and mathematical equations to represent their interrelationships, and to set up the whole system on a computer.

In the first section the relationship $y = \theta x + e$ was introduced as a model of how two particular variables relate to each other. It can be seen from this example that the term model is here used very broadly to describe mathematical and statistical relationships.

A mathematical and/or statistical expression can be used as a model when some of its properties, its behaviour, are similar to the properties, the behaviour, of the quantities being modelled. The benefits of a mathematically presented model are that it is clearly defined and thus easily communicated, so that its strengths and weaknesses may be analysed. It can be manipulated according to the clear rules of mathematics and investigated either analytically or, more experimentally, using a computer. There exist some cases where a situation is so precise and clear that a purely mathematical model can be devised that is accepted as true within clearly defined conditions. However, in most circumstances we lack clear knowledge of the underlying mechanisms and relationships in the situation, our data is limited in coverage and subject to error. We are thus faced with the possibility, if not certainty, that the model we finally devise will be wrong. It may be wrong in form, it may be inaccurate in the numerical values we use in it, it may be inappropriate in some new situation to which we apply it. To properly understand the effects of error in our models we require the methods of statistics. It is the application of these methods in the process of modelling that is the subject matter of this book.

It should now be clear that we cannot expect our models to be true descriptions of reality. A particular model may give a good description of one aspect of the situation but a poor one of another. Our toy aeroplane may look like a Boeing but it will not fly, or it may fly but look nothing like a real plane. Our models are but approximations to certain aspects of a complex reality. If a model is only a limited, and possibly distorted, picture of reality, in what sense can we talk about the truth of a model? The Greek word for truth is *aletheia* which means 'unhiddenness'. A model can be seen as truth insofar as it makes 'unhidden' aspects of the situation being modelled that were previously hidden. Another ancient definition of truth speaks of 'adaequatio intellectus.' Thus the truth of the model does not depend on it being 'the truth, the whole truth, and nothing but the truth' but simply on whether it is adequate, *adaequatio*, to reveal some aspects of reality to our insight. We thus have to judge it not in terms of right or wrong but in terms of insight and use.

We have considered modelling as a means of creating representations of the 'real world'. Obviously this can only be done by focusing on some specific aspect of reality. This subset of reality will consist of a particular object, situation, event, system, etc. It is convenient to have a name for the subset that we intend to model. The term *Prototype* will be used. The prototype may initially be rather vaguely defined, but it is important to seek for a clear specification as the modelling exercise progresses. The following are examples of simple *Prototype Specifications*.

Example 1
The prototype is a sagebrush ecology on the North American continent at altitude between sea level and 3000 metres. There may be similar ecologies in other places but the model does not seek to model those.

Example 2
The prototype is the chemical reaction, and in particular the reaction time, of chemical salts when added to water. The prototype here covers the industrially produced chemical in an ordinary range of United Kingdom tap waters. The prototype based on pure chemical and distilled water is a different prototype.

In defining the problem we thus break 'reality' into the prototype and the rest. For ease of description we shall call the remainder the *Environment* of the prototype. The situation is thus as represented in Figure 1.2.

Most parts of the environment will be totally irrelevant to the problem in hand. The distance of Jupiter is of no relevance to an economic analysis of the steel industry in the United Kingdom. However, in defining the

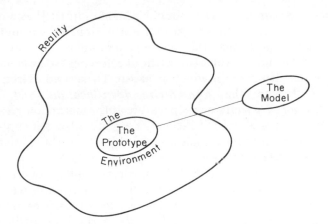

Figure 1.2 The model and reality

prototype many economic factors will also have to be left out to keep the problem to a reasonable magnitude. These might be regarded as the close environment. We do not need to define the environment precisely but we need to be conscious of its presence and of our implicit assumption that the prototype can be sensibly modelled whilst ignoring everything in the environment. During a modelling exercise it often happens that factors in the environment are found to be important and are brought in to the prototype and, conversely, factors in the prototype are not used.

The use of the term 'prototype' might seem to be a misnomer as it refers usually to a first model. It has been used for two reasons: firstly there is a precedent in the literature on mathematical modelling, e.g. the book by Aris. Secondly it underlines that in the prototype specification and the separation of prototype from environment we are already using our conceptual and mental models of the situation. Our very selection involves assumptions, intuitions, perceptions, intentions, that are acknowledged by speaking of the prototype rather than of reality.

1.3 The major stages in statistical modelling

The situation in section 1.1 illustrates that there are five major stages in the process of statistical modelling. These are:

(a) *Identification.* This is the process of finding or choosing an appropriate model for a given situation. The example of section 1.1 indicated that there was no unique way of doing this. Part III of this book discusses some of the considerations and methodologies of value in identification. The study of approaches to identification is still in its early days; there are no rigorous procedures that guarantee success. The example illustrated two totally different and extreme approaches: one a conceptual approach that sought a model on the basis of rational argument from some knowledge of the situation, but without reference to any actual data; the other an empirical approach that considered only the data and its properties without reference to the meaning of the data or the situation in which it arose. These two extremes will be referred to as the *Conceptual Identification* and *Empirical Identification*. We shall look separately at the methodologies that each provide for model identification. We shall also examine how the modeller can, in practice, use and combine both these approaches in *Eclectic Identification*.

(b) *Estimation and fitting*. As we saw in the example, though the general form of a model will be of interest, to use it in practice we must put it into a detailed numerical form. The $y = \theta x + e$ must become $y = 1.3\,x + e$. Clearly we might have used $y = 2x$ as the line but if this line is drawn on Figure 1.1 it is clear that it does not model the data that we have. The line $y = 1.3\,x$ is much closer to the data. This stage of moving from the general model form to the specific numerical form is called *Model Fitting*. The process of assigning numerical values to parameters is called *Estimation*. Part IV of this book looks at a wide range of approaches to this stage.

(c) *Validation*. We shall use the term validation in a very wide sense to cover all those processes that take place when considering the practical value and validity of the model in a given situation. This validation takes place in three stages: in the development stage of the model, when a fitted model provides a yardstick that can be used to reveal further aspects of the data; in the testing stage, when new data is gathered to provide a further validation of the model; in the application stage, when monitoring procedures are introduced to check whether an initially satisfactory model remains valid in use. The methods of validation are discussed in Part V.

(d) *Application*. The application of the models derived using statistical modelling is the subject matter of books on physics, psychology, ergonomics, or whatever branch of human study gave rise to the modelling process. It is therefore not the direct subject matter of this book. However, as we shall see, we cannot divorce the modelling process from the ultimate purpose for which a model is required. We therefore need to look at the interaction between model identification, fitting, and validation and the intended applications of the model. This study is the focus of Part VI.

(e) *Iteration*. A book must be written with its topics in sequence. It must not be thought, however, that the modeller proceeds in the same ordered fashion through the sequence of identification, fitting, validation, and application. Figure 1.3 gives a more realistic view: the process is a process of continuous development, of going back a stage or two to make use of additional information. The model is never THE

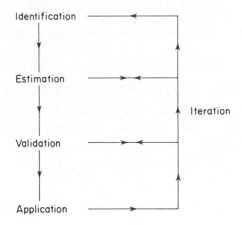

Figure 1.3 The process of statistical modelling

model, final and unalterable; it is always a tentative model that we shall use until we can improve it. Often the stages of the modelling process merge: fitting is done as part of identification; a validation study leads immediately to a newly identified and better model; when fitting a model doubt is cast on its validity. So that this fundamental aspect is not lost, in an ordered presentation, the final part of the book examines in more detail how we may develop models. Our first investigation will be based on the simplest models tenable. As we iterate the various possible loops of Figure 1.3, iteration being any processes of repetition and improvement, then we may need some of the more sophisticated ideas and techniques of this final part.

1.4 Sources of uncertainty in modelling

We have avoided seeking formal definitions of what a model is, we will similarly avoid defining 'statistics'. For our purposes it is sufficient to say that it is a growing range of ideas and methodologies that are concerned with the analysis of data and the construction of models. Their common basis is that they allow for the ever-present uncertainty in our world.

The most difficult source of uncertainty to deal with is that due to the possibility of error in the identification process. Let us refer to this as *Identification Uncertainty*. For example, suppose we seek to model a variable y by using a related variable x and an error term e and develop a model defined by

$$y = \alpha + \beta x + e, \quad \text{where } \alpha \text{ and } \beta \text{ are constants.}$$

This model will clearly be less than satisfactory if in fact y is also significantly affected by a further variable z that is unrelated to x. In looking at past data the effects of z will be contained in the error term, so some allowance is made for the possibility of such unknown factors. However, if we are seeking to use the relationship, the neglect of z may have undesirable effects. A similar situation might occur if a better model would allow for x to influence y in a more complex fashion, for example

$$y = \alpha + \beta x + \gamma x^2 + e, \qquad \text{where } \alpha, \beta \text{ and } \gamma \text{ are constants.}$$

We need always to bear in mind that our model is always a model, it is not the reality. Thus in a sense all models involve identification uncertainty. With simple, well-understood situations the uncertainty may be small and need not worry us. With more complex and ill-understood situations we need to have considerable scepticism about the models we use. In this case we need to explore carefully the possible consequences of identification uncertainty, this we will do in a number of later sections, particularly section 16.7.

A particular aspect of *Identification Uncertainty* relates to the validation aspect of modelling. We live in a world whose uncertainty often arises from its instability. Except in the physical sciences there is no guarantee that the reality being modelled will remain constant in time. Therefore a model that initially has little uncertainty will become more uncertain as time goes by. This is the underlying motive for the methods of monitoring models that we discuss in section 17.5.

We shall often find ourselves examining how the model compares with or relates to the real world. By the real world what we actually mean is the prototype, that very limited sub-set of reality on which we are focusing our attention. So, not only may we have uncertainty because we have the wrong model for the right prototype, we may also have the wrong prototype for the application.

The other major source of uncertainty is our data. When we seek to fit models to our data we shall begin to see how this uncertainty occurs. We shall often find that an accurate fit requires more data than we can either practically or financially obtain. Even where we have enough data we will find that its quality is often far from perfect. There will not only be the variability in the data due to the natural variability of the quantities being measured, there will often be an 'unnatural' variability due to mistakes in measurement, transcription, misunderstanding. We shall have accurate data on some variables, data on other variables will be inadequate. Some quantities are clearly defined and accurately measurable, others are general ideas that are represented in the model by measurements on other factors that act as proxies. In our study we shall examine methods of detecting and allowing for the gross faults in data. We shall also consider in greater depth the consequences of limitations on the amount and quality of data available. At the centre of this consideration is the idea of *Sampling Variability*. Almost any set of data we possess is a sample from a conceivably infinite set of possible observations. To obtain an exact fit of our model we would require an infinitely large sample. This imagined set of all possible observations we refer to as the *Population*. If we were to take a large sample we would hope that the characteristics of the population were reflected in the characteristics of the sample. As the sample gets smaller this hope gets more speculative. For example, suppose we consider a population that represents the errors occurring in measuring some physical constant. A sensible model might assume that the population of errors is symmetrically spread about zero. A sample of only two measurements, however, has a 50 : 50 chance of having both errors with the same sign. It is only by taking larger samples that we can hope to get an average measurement that is satisfactorily close to the true value. The effect of sampling variation is not only to reduce the precision with which we can fit models but also to limit the precision with which the model may then be used. The existence of sampling variability means that the future samples will vary from past samples and hence from the model fitted using past samples. Sampling variation thus causes uncertainty at all stages of modelling.

Chapter 2

Model structures

2.1 Introduction

In essence a model is something whose structure, and hence behaviour, corresponds in some sense to that of a particular reality or phenomenon. In statistical modelling the structures that we shall use will usually be mathematical in notation and will have elements of chance in their nature. The underlying process may itself be completely deterministic, but if our evidence or perception of it is uncertain then statistical aspects of modelling become relevant. The objective of this chapter is to introduce the basic components that make up the structures we shall be studying and to give some simple illustrations of their use in models.

2.2 Variables and parameters—an example

Consider a model of sales that show a steady upward trend. In words we could write

sales in week t = level of trend in week t + chance variation,

and in symbols

$$x_t = \alpha + \beta t + e_t,$$

where $\alpha + \beta t$ is a trend in sales, with a weekly change of β units, a level α in week zero and e_t represents the chance variation, from this trend, in week t. Such a mathematically formulated model is designed to represent the way in which the various factors, sales, time and chance variation, in the situation relate to each other. We refer to these factors of the *Variables*, of the situation and the model, since our interest is mostly in studying the ways in which these quanatities vary. In this model there are three variables t, e_t and x_t. The variable t is known and determined in a given week and is referrred to as a *Determined* or *Deterministic Variable*. If we want to forecast sales for week 14 we know that $t = 14$ in that week. However e_{14} is, until week 14, an unknown and random quantity. We may, however, be able to make probability statements about it. We might be able to say that, on past evidence, there is a low probability, say 1/100, of e_{14} being outside the range -200 to $+200$ units. A variable with this type of behaviour we call a *Random Variable*. The modelled variable x_t is thus a variable whose behaviour has both deterministic and random elements. In this model we are using the variables on the right side of the equation to explain the behaviour of the variable on the left side. This left-hand variable is said to be the *Dependent Variable* which depends on the values of the variable on the right side which is called the *Independent Variable* or *Regressor Variable*. We shall use the term regressor variables to avoid later complications of referring to interdependent independent variables. The model also contains two quantities α and β that determine the magnitude of the trend and which are not directly measurable. These are referred to as the *Parameters* of the model. There may also be other, hidden, parameters in the model, for the probabilities associated with e_t may depend on other parameters, say μ and σ. It will be seen that the parameters, α, β, μ, σ, will determine the numerical values of x_t for each t, at least to within probabilistic limits. The parameters thus control the magnitude of the effects that the regressor variable has on the dependent variable. By convention we denote variables by Latin characters X, Y, x, y and parameters by Greek characters α, β, μ, ...

An important classification of variables, both deterministic and random, is based on the values that the variable can take. The week number t is based on a counting process and can only take integer values 0, 1, 2, 3, ... A variable that can only take one of a number of specified values is called a *Discrete Variable*. If the variable can take any value on a continuous axis, then it is called a *Continuous Variable*. If, in the sales example, sales refers to the length of material sold then this would be truly continuous. If the sales referred to the number of reels of thread sold this by its nature would be formally discrete. However, if we sold many thousands so $x_{12} = 11{,}295$, $x_{13} = 12{,}483$, then for all practical purposes x_t, and e_t, may be treated as continuous variables.

If we consider how variables are measured we obtain a further useful classification. At the most primitive level of measurement we simply classify things. The essays are marked A, B, C; the insect is classified as a Leaf Hopper or a Rove Beetle. Data in this original form is called *Nominal* data. In practice we usually convert to numbers by counting the numbers in each class. At the next level of measurement we order the items being measured. We order a set of essays, or noise environments on the basis of some general criteria. Such rank orderings are referred to as *Ordinal* data. Finally we have data where we measure against some scale and get measurements of 20°C, 6 km etc. Such data is said to be measured on an *Interval/Ratio Scale*. To be on a Ratio Scale the variable must have a meaningful zero, thus 6 km is twice 3 km so km is a Ratio Scale. However, 20°C is not twice as hot as 10°C so °C is only an Interval Scale.

2.3 Deterministic variables

Variables that can be treated as deterministic arise in a number of ways:

(a) *Variables that are under direct and accurate control*
A model for sales may contain the variable $n =$ the number of sales outlets. This value will be precisely known and indeed is controllable. We might consider as a somewhat different example a model of an experiment in which a variable is the amount of fertilizer applied per acre. This might be assigned in the experiment at 1, 2, and 3 kg on three different plots. There will be some randomness in this since the actual amount will in practice vary from exactly 1, 2, or 3. However, in relation to the other sources of variability in the experiment, such as local variations in soil fertility, the random variability in total fertilizer added to each experimental plot may be neglected and the variable treated as deterministic.

(b) *Variables used retrospectively*
For example the population, P, of a town in 1982 when viewed from 1980 is a random variable. However, in 1983 P is known to be 69,300 and in modelling the economy of the town P can then be treated as a deterministic varibale. Even in 1980 we could still have done such modelling making the assumption that the population would be 70,000. Again P would be treated as a deterministic variable. In this case we would refer to the modelling as being *conditional* on P being 70,000.

2.4 Random variables

In this section we shall assume the reader understands the basic ideas and rules of probability. Appendix 1 gives a brief introduction to the main ideas.

With deterministic variables it is possible to write $x = 6$ and be totally clear as to what this statement means. With random variables this is not the case. Before x is observed it is not known what value it will take though it may be possible to make some probability statements about it. After x is observed as having the value 6 we can still only attach significance to that value against a background framework of probability statements. Thus in discussing the concept of a random variable we must look in some detail at these probability statements.

It is useful to introduce a notation of capital and small letters to distinguish between the random variable, X, and its numerical value on any specific occasion, which can take on one, x, of a possibly infinite number of values. We cannot say what value X will take but only assign probabilities to the various possible values of x. The value x is sometimes called the *Realization* of X.

Example 1

A vending machine usually provides coffee on insertion of a coin. Over a period of time it is found not to work on two 'uses' in every thousand. Suppose we use X as an 'indicator variable' so that $x = 1$ if the machine works and $x = 0$ if no coffee is obtained. Thus the model is

$$\text{Probability machine works} = 0.998$$
$$\text{Probability machine does not work} = 0.002$$

or

$$Pr(X = 1) = 0.998$$
$$Pr(X = 0) = 0.002$$

or

$$Pr(x) = \begin{array}{ll} 0.998 & x = 1 \\ 0.002 & x = 0, \end{array}$$

where $Pr(\quad)$ denotes the probability of the event $(\quad)$.

A variable X for which such a *Probability Distribution* is required to describe its random behaviour is hence called a random variable. If we make this model completely general to cover any level of machine reliability we could express it as

$$Pr(X = 1) = \theta$$
$$Pr(X = 0) = 1 - \theta \qquad 0 \leqq \theta \leqq 1$$

where θ, the probability of obtaining coffee, is the parameter of the distribution of X.

It is seen from this example that a model for a situation of randomness describes the probability structure by assigning probabilities to values of a random variable, by showing how the total probability is distributed over the possible values taken by X.

A random variable can take on either discrete values, for example $R = 0, 1, 2, \ldots , 10$, or continuous values, for example x may be any value in the range zero to infinity $(0, \infty)$. For a discrete random variable R the distribution is defined by the formula that gives all the separate probabilities of the possible observations r.

$$Pr(R = r) = P(r), \qquad r = \text{each possible value}$$

As one of these values *must* occur their sum (denoted by $\sum$) is one, i.e.

$$\sum_{\substack{\text{all values} \\ \text{of } r}} P(r) = 1,$$

since a certain event has probability 1. Similarly, by the rules of probability (Appendix 1) the probability that R takes on one or other of several discrete values $u, v, \ldots , w$ is given by

$$Pr(R = u \text{ or } v \text{ or } \ldots \text{ or } w) = P(u) + P(v) + \ldots + P(w).$$

Example 2

Repeated attempts are made to complete a task within a given time. If no expertise develops with practice the probability, θ, of succeeding at each attempt remains constant. If the trials are independent the rules of probability show that the probability of getting the first success on the nth trial is

$$Pr(N = n) = Pr(\text{fail on 1st and 2nd and} \dots \text{and } n-1\text{th and}$$

$$\text{succeed on the } n\text{th})$$

$$= (1-\theta)(1-\theta) \dots (1-\theta)\theta$$

$$= (1-\theta)^{n-1}\theta$$

where $1 - \theta$ is the probability of failure and N is the random variable being considered, which can take the values $n = 1, 2, 3, \dots$, without limit. The probability function $Pr(N = n)$, called the *Geometric Distribution*, is discussed further in section 3.2 and Figure 3.2 shows its shape. It is a property of series that if $0 < a < 1$, a being some constant,

$$1 + a + a^2 + a^3 + \dots \text{ to infinity} = \frac{1}{1-a}.$$

It follows by putting $a = 1 - \theta$ that

$$\sum_{n=1}^{\infty} Pr(N = n) = \theta + a\theta + a^2\theta + \dots = \theta\frac{1}{1-(1-\theta)} = 1$$

Further

$$1 + a + \dots + a^m = \frac{1 - a^{m+1}}{1-a}$$

So

$$Pr(N \leqslant n) = \theta \sum_{s=1}^{n} a^{s-1} = \theta\frac{1 - (1-\theta)^n}{1-(1-\theta)}$$

$$= 1 - (1-\theta)^n.$$

This is referred to as the *Cumulative Distribution Function (c.d.f.)* for the Geometric Distribution.

If X is a continuous random variable there are of necessity an infinity of possible values that X can take. Thus the probability that X takes on any specific value will be zero. We thus have to relate probabilites to ranges of values of X rather than to single values. We can do this from two possible starting points

(a) We can define the structure by

$$Pr(X \leqslant x) = F(x).$$

$F(x)$ is the cumulative distribution function. Figure 2.1(a) shows the general form of a c.d.f. As X is certain to take some value less than infinity $F(\infty) = 1$ and as it is impossible to be less than $-\infty$, $F(-\infty) = 0$.

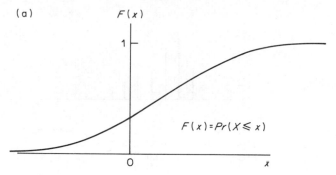

A cumulative distribution function

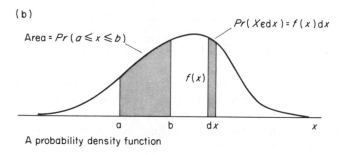

A probability density function

Figure 2.1 The shape and interpretation of c.d.f. and p.d.f.

(b) We can define the structure by taking an infinitesimal range, denoted by dx, of x, i.e. $(x, x+dx)$, and considering

$$Pr(X \text{ lies in } (x, x+dx)) = f(x)\,dx$$

The function $f(x)$ is called the *Probability Density Function* (p.d.f.). It can be seen from the figure and the above definition that the probability $f(x)\,dx$ is the area of the infinitesimal region and that in general the probability that X lies in any range, say (a, b) in the figure, is the area under the p.d.f. in that range. In the discrete case $P(r)$ is the quantity that is analogous to $f(x)$, or to be precise to $f(x)\,dx$. The c.d.f. would be given by

$$F(r) = \sum_{\text{all } R \leqslant r} P(R).$$

The relation between $F(x)$ and $f(x)$ for continuous x is apparent from Figure 2.1(b) since the area under $f(x)$ to the left of $X = x$ is the probability that $X \leqslant x$ which is, by definition, $F(x)$. Furthermore the area between $X = a$ and $X = b$ could be found by $F(b) - F(a)$ which is also the probability that X lies in (a, b). We will usually define a distribution in terms of its

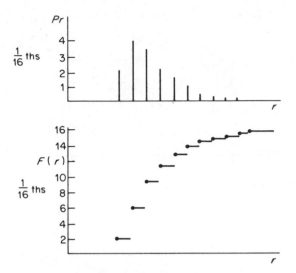

Figure 2.2 The discrete p.d.f. and c.d.f.

p.d.f., $f(x)$, but for actually evaluating probabilities the c.d.f., $F(x)$, is most convenient. Sometimes $F(x)$ is derived from $f(x)$ using mathematical methods. Where this is mathematically difficult tables of $F(x)$ are often available. The shape of the p.d.f. and c.d.f. for a discrete random variable is illustrated in Figure 2.2.

Example 3
Loadings greater than a certain amount cause a component of a production line to temporarily stop. Such loadings occur at random with a mean time between them of μ time units ($\mu \geqslant 0$). For such a situation the distribution of the time between stoppages, T, often has a probability density function

$$f(t) = \begin{array}{ll} \dfrac{1}{\mu}\,e^{-t/\mu} & t \geqslant 0 \\[2mm] 0 & t < 0 \end{array}$$

This model is called the *Exponential Distribution*, shown in Figure 2.3. Thus T is a random variable which can only take non-negative values. The probability that T lies in the infinitesimal interval $(t, t + dt)$ is $f(t)\,dt$. The probability that $T \geqslant t$ is the area to the right of $T = t$ which can be shown to be $e^{-t/\mu}$. Thus the probability that $T \leqslant t$, which is $F(t)$, is

$$F(t) = 1 - e^{-t/\mu}, \qquad t \geqslant 0.$$

For example the probability of a stoppage occurring in less than the time of $t = \mu$ is

$$F(1) = 1 - e^{-1} = 0.63$$

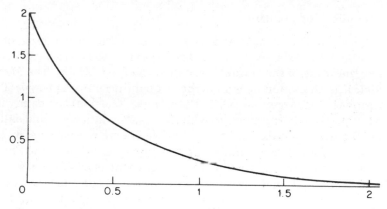

Figure 2.3 The p.d.f. of the exponential distribution ($\lambda = 2$) $\mu = \frac{1}{2}$

The probability of working for more than three times the time $t = \mu$ is

$$1 - F(3) = e^{-3} = 0.05$$

It is always helpful in choosing models for particular situations to have a list of suitable models classified in various ways. There are three separate considerations that are helpful in classifying distributional models:

(a) the nature of the random variable,
(b) the meaning of the parameters,
(c) the shape of the distribution.

In the exponential model of the last example the time to stoppage must be (i) a continuous random variable and (ii) a positive variable. These two facts narrow the possibilities considerably. The mean time between stoppages can be shown to be μ. Thus the parameter of the model has clear significance in the situation being modelled. A choice of model whose parameters are readily interpretable is always an advantage. Finally the shape of the distribution needs to correspond to the form shown by available data and to the intuitive form suggested by the situation. In the example one would expect the p.d.f. to die away to zero as t gets larger as in Figure 2.3.

2.5 Summarizing properties of distributions

With distributional models the shape of the distribution is not necessarily obvious from the formula of the probability density function. It is therefore useful to define some quantities that indicate the shape of the distribution. We shall not go into the mathematical formulation of these measures but simply indicate their nature and general use.

2.5.1 Measures of position

There are a number of measures that indicate the positioning of the distribution on the axis, see Figure 2.4. The peak of the distribution curve occurs at the value of the variable, x, called the *Mode Mo(x)*. The *Median* of x, *Me(x)*, is the value at which the probabilities are split $50:50$ for observations being larger or smaller than *Me(x)*. The final commonly used measure of position corresponds to what for a set of data we would call the average. We will call this the *Population Mean* or *Expectation* of x, $E(x)$. This is the average of the infinite population of values described by the distribution. The appropriateness of these three measures depends on the use being proposed and the distribution being considered. For the exponential distribution of stopping times T for example the values are

$$Mo(T) = 0$$
$$Me(T) = \mu \ln 2 \text{ (where ln is the natural logarithm)}$$
$$E(T) = \mu$$

Thus in this case the parameter μ corresponds to the mean of the population of possible stopping times T.

We shall be making substantial use of the idea of expectation and in section 5.2 some of the rules governing its behaviour will be given. The actual calculation of expectations often involves a level of mathematics beyond that set in this book, so the interested reader is referred to any of the more mathematical texts in statistics, e.g. Freund and Walpole (see References and Reading list, Part II). We will limit ourselves to two simple illustrations of the basic idea for discrete distributions.

Example 1
Suppose in the coffee machine example we want the expected number of failures per 10,000 cups. The probability of failure is $(1 - \theta)$ for a single cup, so it is intuitively reasonable to expect $10,000 (1 - \theta)$ failures for 10,000 cups. This assumes the trials are independent of each other and that θ remains constant.

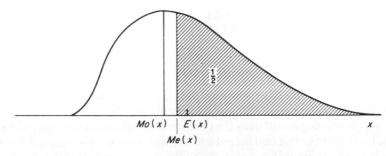

Figure 2.4 Mean, median, and mode of a distribution

Example 2

Consider again the Geometric Model. Suppose we carry out a large number, M, of trials getting first successes at the $n_1, n_2, \ldots n_M$ trials. The average of these ns is

$$\frac{n_1 + n_2 + \ldots + n_M}{M}$$

Without such experimental evidence we could argue that, as the probability of a specific value of n is given by

$$Prob(N = n) = (1 - \theta)^{n-1}\theta,$$

the expected number of observations of first success at n would be $M(1-\theta)^{n-1}\theta$. Thus for M sufficiently large the expected data would consist of

$$\begin{array}{ll} n = 1 & M\theta \text{ times} \\ n = 2 & M(1-\theta)\theta \text{ times} \\ n = 3 & M(1-\theta)^2\theta \text{ times} \\ & \text{etc.} \end{array}$$

The average of this expected data will give the expectation of the random variable N as

$$E(N) = \frac{1 . M\theta + 2 . M(1-\theta)\theta + 3 . M(1-\theta)^2\theta + \ldots}{M}$$

$$= \theta \, [1 + 2(1-\theta) + 3(1-\theta)^2 + \ldots].$$

A standard algebraic result is that

$$1 + 2a + 3a^2 + 4a^3 + \ldots \text{ to infinity} = 1/(1-a)^2$$

It follows that for the Geometric Distribution the expectation is

$$E(N) = \theta/[1 - (1-\theta)]^2 = 1/\theta.$$

As intuitively justified in the example the general form for the expectation of a discrete random variable R is:

$$E(R) = \sum_{\text{all values}} r \, Pr(R = r)$$

We shall sometimes be interested in the expected value not of the random variable itself but of some variable derived from it, z_r, say. We might have $z_r = r^2$ or $z_r = $ profit obtained when $R = r$. We can use the same arguments to justify the formula

$$E(Z_r) = \sum_{\substack{\text{all values} \\ \text{or } r}} z_r \, Pr(R = r)$$

or alternatively treat z as a new random variable, find its distribution and write

$$E(Z) = \sum_{\substack{\text{all values} \\ \text{of } z}} z \, Pr(Z = z).$$

Example 3

People are tested on the trials of the previous example and 'selected' if they succeed first trial. Let $X = 1$ if a person succeeds first trial and $X = 0$ otherwise. Assume all the people have the same probability of success, θ. Then we can argue that

$$E(X) = \sum_{n=1}^{\infty} x \, Pr(N = n)$$
$$= 1.\theta + 0.(1-\theta)\theta + 0.(1-\theta)^2\theta + \ldots$$
$$= \theta,$$

or that

$$X = 1 \text{ wth probability } \theta$$
$$X = 0 \text{ with probability } 1-\theta$$

so

$$E(X) = 1.\theta + 0.(1-\theta)$$
$$= \theta.$$

2.5.2 Measures of spread

Most measures of spread are based on considering the deviations of the values in the population from their expected value, $E(X)$. Thus the deviation

$$D = X - E(X)$$

is considered. To consider the magnitude of D without being concerned with its sign we may use its absolute value, written as, $|D|$ or its square D^2. If the expectation, the population average, of these are defined we have

$$MAD(X) = E(|D|)$$

and

$$V(X) = E(D^2)$$

as two measures of the variability in X, called the *Mean Absolute Deviation* and the *Variance* respectively. If the distribution has a wide spread these quantities will be relatively large. If X tends to concentrate close to the mean they will be relatively small. As $V(X)$ has dimensions of x^2 we often use its square root called the *Standard Deviation* and denoted $\sigma(X)$.

2.5.3 A measure of skewness

If a distribution is symmetric in shape then $Mo(X) = Me(X) = E(X)$ and D will have equal probability of being positive or negative. If the distribution is not symmetric it will have a longer tail either to the right or the left as in Figure 2.5. If the tail is to the right we say the distribution is *Positively Skewed*, if to the left *Negatively Skewed*. A measure of skewness is provided

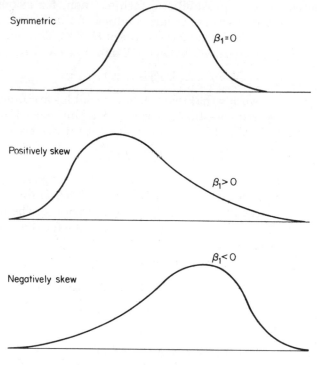

Figure 2.5 Skew distributions

by the population mean of D^3, i.e. by

$$\mu_3 = E(D^3).$$

Writing $E(D^2) = \mu_2$ a standardized measure of skewness is the coefficient of skewness β_1 where

$$\beta_1^2 = \mu_3^2/\mu_2^3$$

with β_1 given the sign of μ_3. A symmetric distribution will give $\mu_3 = \beta_1 = 0$. For the exponential distribution $\beta_1 = 2$. The quantity $\mu_r = E(D^r)$ is called the rth Population *Moment about the Mean*. Derivations of all these various measures of distributional shape are given in most standard statistical texts. A table giving these values for some common distributional models is given in Table 3.2 of Chapter 3.

2.5.4 Measures of relationship

The concept of expectation can be used to develop a measure of linear relationship between two random variables, X, Y. If there is no linear relationship we would assume that the expectation of XY would simply be the product of the expectations of X and Y. If, however X and Y were linearly related in some probabilistic fashion then, for example, large values of X would tend to relate to large values of Y and vice versa. This would have the effect of increasing the value of $E(XY)$. The amount of this increase defines the *Covariance* between X and Y, $C(X, Y)$:

$$C(X, Y) = E(XY) - E(X)E(Y).$$

In probability terms we say that two random variables are *Independent* if the probability properties of either one is in no way influenced by the value taken by the other. If X and Y are independent it intuitively follows that

$$E(XY) = E(X)E(Y)$$

and hence that $C(X, Y) = 0$. The converse is not in general true. For example if we have a non-linear relation such as the points (X, Y) lying randomly round a circle, then $C(X, Y) = 0$ in spite of there being a strong relation between X and Y. Returning to the variance for the moment note that it can be written as

$$V(X) = E[(X - E(X))^2]$$
$$= E[X^2 - 2E(X)X + E(X)^2].$$

The expectation of any non-random constant, such as 2 or $E(X)$, is just the value itself so this expression simplifies to

$$V(X) = E(X^2) - E(X)^2,$$

which is the same as the covariance of X with itself. This explains the term covariance and it also suggests the following standardization of $C(X, Y)$. We define the *Correlation* between X and Y by

$$\rho(X, Y) = \frac{C(X, Y)}{\sqrt{V(X)V(Y)}}$$

If $Y = X$ then $\rho = 1$ and if $Y = -X$ then $\rho = -1$. If Y and X are independent then $\rho = 0$. It can be shown that ρ lies in the interval $(-1, 1)$. The correlation between X and Y is thus a measure of linear relationship that has numerical values that are easily interpretable, the relationship being stronger the closer ρ is to -1 or $+1$.

A reasonable requirement of a measure of linear relationship is that artificial changes in origin and scale for the variables should not change the value. This is not true for $C(X, Y)$ but it is true for $\rho(X, Y)$. So for example

$$\rho(2X + 3, 5Y - 7) = \rho(X, Y).$$

2.6 The likelihood

In the previous discussion of distributional models we have considered two particular simple examples:

(a) The, discrete, geometric distribution

$$Prob(N = n|\theta) = (1-\theta)^{n-1}\theta, \qquad n = 1, 2, \ldots.$$

(b) The, continuous, exponential distribution with parameter $\lambda = \dfrac{1}{\mu}$, to simplify the notation from our previous use,

$$f(x|\lambda) = \lambda e^{-\lambda x}, \qquad x \geqslant 0.$$

So far these formulae have been looked at as functions of a random variable, n or x, of use in calculating probabilities and it has been assumed that the parameters, θ and λ, are known. We have emphasized this by writing (random variable|parameter) to indicate that the parameter is assumed known. Thus for (a) we would refer to the 'probability that $N = n$ given θ'. If the data consisted of independent observations $N = 3, 5, 4, 5$ we could write

$$Prob(3, 5, 4, 5|\theta) = [(1-\theta)^{3-1}\theta]\,[(1-\theta)^{5-1}\theta]\,[(1-\theta)^{4-1}\theta]\,[(1-\theta)^{5-1}\theta]$$
$$= (1-\theta)^{13}\theta^4.$$

This will give us the exact probability for any given θ. Figure 2.6(a) shows this probability as a function of θ. It will be seen that for $\theta = 0.5$ and larger, the probability of obtaining this data is relatively small, the data is most likely to occur for θ in the region of $\theta = 0.235$. Plotting the probability this way is a useful technique in modelling since we are frequently concerned with comparing the likely occurrence of the data for different parameter values. Indeed this approach is sufficiently important for us to use a

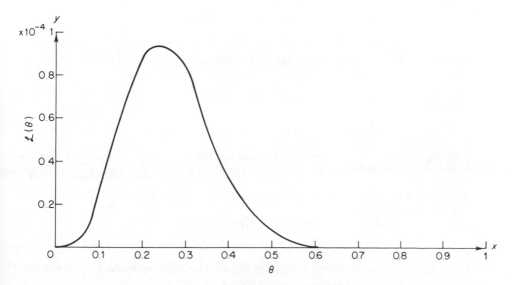

Figure 2.6 The likelihood of the geometric data

different terminology. The function $(1-\theta)^{13}\theta^4$, is called the *Likelihood Function*

$$\mathscr{L}(\theta) = (1-\theta)^{13}\theta^4, \qquad 0 \leqslant \theta \leqslant 1.$$

Observe that the likelihood is the probability of obtaining the given data expressed as a function of the parameter θ. In this situation probability and likelihood are effectively the same thing viewed from different directions, as is symbolized in Figure 2.7.

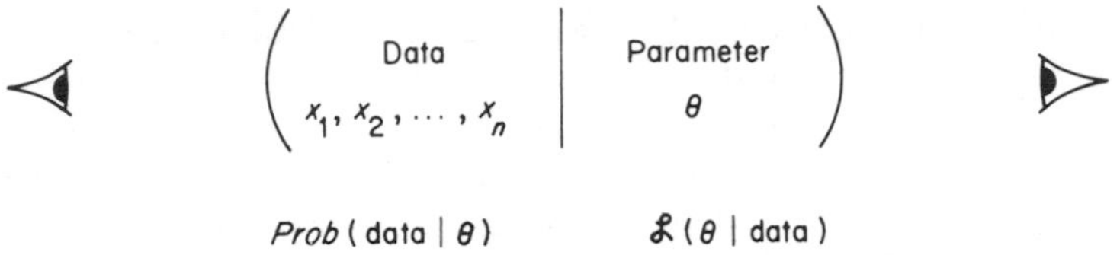

Figure 2.7 Probability and likelihood

We shall frequently make use of the likelihood function by comparing its magnitude at different values of θ. In this, the most common use, the existence of multiplying constants has no effect on the comparisons so it is normal practice to drop such constants in writing the likelihood.

Example 1

Suppose that in the sequence of 3, 5, 4, 5 trials of the first example there had been one success only in each set of trials, but it had not necessarily been on the last trial. The probability would then be

$$(3 \times 5 \times 4 \times 5)(1-\theta)^{13}\theta^4 = 300(1-\theta)^{13}\theta^4,$$

since, for example, in the three trials the single success could have been in any one of these trials. However, for comparison of different values of θ, the 300 is irrelevant so we would write the likelihood as

$$\mathscr{L}(\theta) = (1-\theta)^{13}\theta^4.$$

It occasionally occurs that we wish to compare probabilities or probability densities from different distributions and still use the perspectives provided by the liklihood idea. In these situations we cannot ignore the constant terms. We will refer to likelihoods without removal of constant terms as *Initial Likelihoods* and denote them by $\mathscr{L}^*(\theta)$. Thus, for the example

$$\mathscr{L}^*(\theta) = 300(1-\theta)^{13}\theta^4.$$

Example 2

In an example in Section 2.4 we considered the time between stoppages to be modelled by the exponential distribution

$$f(t|\lambda) = \lambda e^{-\lambda t}, \qquad t \geqslant 0.$$

This is the probability density function and is not a probability. To get a probability we would have to consider the probability of stoppages occurring in an interval dt round t giving

$$Pr(T \text{ in } dt) = f(t|\lambda)dt.$$

If the observed times are 0.8, 1.2, 1.4 then

$$Pr(\text{data in intervals } dt_1, dt_2, dt_3 | \lambda)$$
$$= \lambda e^{-\lambda 0.8}dt_1 . \lambda e^{-\lambda 1.2}dt_2 . \lambda e^{-\lambda 1.4}dt_3,$$
$$= \lambda^3 e^{-3.4\lambda}dt_1 dt_2 dt_3.$$

Plotting the shape of this gives Figure 2.8. It will be seen that the probability of getting data in the three intervals is much greater with $\lambda = 1$ than for example with $\lambda = 0.1$ or $\lambda = 10$. The value $\lambda = 1$ thus seems more plausible than, more likely than, $\lambda = 0.1$ or $\lambda = 10$. Notice that the relative heights on the plot do not depend on the intervals dt_1, dt_2, dt_3 but only on the function $\lambda^3 e^{-3.4\lambda}$. Thus in this situation we define the likelihood as equal to this function and hence only proportional to the probability. As it is proportional to, symbolized by $\propto$, the probability density, we need not worry about the distinction between probability densities and probabilities. Thus

$$\mathscr{L}(\lambda|\text{data}) \propto f(\text{data}|\lambda)$$

so

$$\mathscr{L}(\lambda|\text{data}) = \lambda^3 e^{-3.4\lambda}.$$

In summary we define the likelihood for any parameter θ given a set of data by

$$\mathscr{L}(\theta) = Prob(\text{data}|\theta)$$
$$= Pr(r_1|\theta)Pr(r_2|\theta) \ldots Pr(r_n|\theta)$$

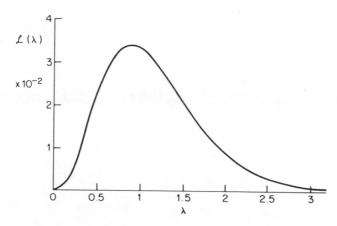

Figure 2.8 The likelihood for exponential data

for a set of independent discrete observations $r_1, r_2, \ldots, r_n$ and by

$$\mathscr{L}(\theta) = f(x_1|\theta)f(x_2|\theta) \ldots f(x_n|\theta)$$

for a set of independent continuous observations $x_1, x_2, \ldots, x_n$, having p.d.f. $f(x|\theta)$.

On the argument presented the likelihood is proportional to the probability of obtaining the data as a function of the parameter. From the plots of likelihood functions shown it is seen that this quantity is high for some values of the parameter and low for others. Some values of the parameter provide a better explanation of the observed data than others or conversely the data provides better support for some parameter values than others. The *Log Likelihood*, $L(\theta)$, sometimes called the *Support*, is defined by

$$L(\theta) = \ln \mathscr{L}(\theta).$$

The maximum of $L(\theta)$ will occur at the same value of θ as that for $\mathscr{L}(\theta)$. The logarithm turns the multiplication of terms like $Pr(r_i|\theta)$, $f(x_i|\theta)$ into the addition of $\ln[Pr(r_i|\theta)]$ or $\ln[f(x_i|\theta)]$, e.g. $L(\theta) = \sum_1^n \ln f(x_i|\theta)$.

Thus the evidence about θ becomes additive over the data. Additional data just adds additional support.

Part II

Common statistical models

Chapter 3

Distributional models

3.1 Introduction

In the previous chapter we introduced the concept of the probability density function as a model of the way a random variable behaves. The p.d.f. describes how probability is distributed over the possible values a random variable can take. Models defined in this fashion are referred to as *Probability Distributions*. Random variables and their distributions are the main elements of some models and components of many others. In this text we are concerned only with models that involve random elements, either inherently or through errors in measurement. Thus the natural starting place for a study of commonly used models is the simple distributional model. In this chapter we illustrate some of the simplest and most useful distributions. The Table 3.3 lists some further distributions, most of which will be used in later chapters. The book by Hastings and Peacock gives a survey of the properties of the most used models.

3.2 The Bernoulli process and related distributions

Consider a sequence of independent trials, labelled 1, 2, 3, ... At each trial an event S either occurs or does not occur. Using conventional terminology we refer to the occurrence of S as a success and its non-occurrence as a failure, F. This terminology describes a vast range of phenomena. For example:

Trial	*Success S*
Shoot at target	Hit target
Sow seed	Seed germinates
Test component	Component works
Student takes examination	Student passes
Test drug sample for safety	Drug safe
Give experimental insecticides to insect	Insect dies
Classify plant	Plant is of variety, V.

A sequence of such events is referred to as a Bernoulli Process.

For a Bernoulli Process we require that the probability of a 'success' at each trial in a sequence, or process, of trials is a constant, θ, independently of the results of other trials. Thus

$$Pr(S) = \theta$$

and

$$Pr(F) = 1 - \theta, \text{ independently for all trials.}$$

Consider just two trials, diagrammatically presented as

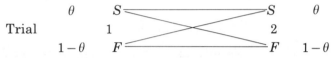

The number of successes, R, may be

$$R = 0, \; Prob(R = 0) = Pr(F \text{ and } F) = Pr(F) \times Pr(F) = (1 - \theta)^2$$

$$R = 1, \; Prob(R = 1) = Pr[(S \text{ and } F) \text{ or } (F \text{ and } S)]$$
$$= [Pr(S) \times Pr(F)] + [Pr(F) \times Pr(S)]$$
$$= 2\theta(1 - \theta)$$

$$R = 2, \; Prob(R = 2) = Pr(S \text{ and } S) = Pr(S) \times Pr(S)$$
$$= \theta^2,$$

making use of the multiplication and addition of probabilities, see Appendix 1. If we generalize this argument for the case where there are n trials with r successes we have

$$R = 0, \; Pr(F \text{ and } F \text{ and } F \ldots F) = (1 - \theta)^n$$

$$R = 1, \; Pr((S \text{ and } F \text{ and } \ldots F) \text{ or } (F \text{ and } S \text{ and } F \ldots F) \text{ or } \ldots$$
$$(F \text{ and } F \ldots FS)) = n\theta(1 - \theta)^{n-1}$$

$$R = r, \; Pr((S \text{ and } S \text{ and } S \ldots F) \text{ or } \ldots \text{ or } (F \ldots S \text{ and } S))$$
$$= K\theta^r(1 - \theta)^{n-r}$$

$$\cdots \cdots \cdots \cdots \cdots \cdots \cdots$$

$$R = n \; Pr(S \text{ and } S \ldots S) = \theta^n$$

where K is the number of ways the r S's can occur in the n trials. It can be shown this number, called the Binomial Coefficient, is

$$\frac{n!}{r!(n-r)!} \text{ where } r! = r(r-1)(r-2) \ldots 3 \times 2 \times 1.$$

This is symbolized by $\binom{n}{r}$ or ${}^nC_r$. For example with $n = 5$ trials and $r = 3$ successes

$$\binom{5}{3} = \frac{5!}{3!2!} = \frac{5 \times 4 \times 3 \times 2 \times 1}{(3 \times 2 \times 1) \times (2 \times 1)} = 10,$$

so there are 10 different sequences of S and F that contain $3S$ and $2F$ in 5 trials.

In general if R is the random variable denoting the number of successes in n trials

$$Pr(R = r) = \binom{n}{r}\theta^r(1 - \theta)^{n-r} \qquad r = 0, 1, \ldots, n.$$

This is called the *Binomial Distribution*. Figure 3.1 illustrates the shapes of this distribution for various values of n and θ. It can be observed that θ controls the symmetry of the distribution, with perfect symmetry when $\theta = \frac{1}{2}$ and a strong skewness when θ is near zero or one. As a symbol for a random variable R to have a Binomial Distribution we will write

$$R \cap \text{Bin}(n, \theta).$$

It has been shown that the Binomial Distribution arises from the Bernoulli Process by considering a fixed number of trials, n, and having the number of successes, R, as the random variable. We may however work the other way round by fixing the number of successes at r and having as the random variable the number of trials, N, required to obtain these r successes. Consider the simplest case of the number of trials, N, required to obtain the first success. The probability that $N = n$ is the probability of $n - 1$ successive failures followed immediately by the one success. The probability of this is

$$Prob(N = n) = (1 - \theta)^{n-1}\theta \qquad n = 1, 2, \ldots.$$

It will be seen that unlike the Binomial the random variable can take any integer value $n = 1, 2, \ldots$. This distribution is the Geometric Distribution derived in Chapter 2. Figure 3.2 indicates its shape for various parameters θ. For the case of r successes in n trials there must be $r - 1$ successes in the first $n - 1$ trials and a success on the nth. Thus the probability distribution is given by

$$Pr(N = n) = \binom{n-1}{r-1}\theta^{r-1}(1 - \theta)^{(n-1)-(r-1)}\theta$$

$$= \binom{n-1}{r-1}\theta^r(1 - \theta)^{n-r}, n = r, r+1, \ldots.$$

This distribution is called the *Pascal Distribution* (Figure 3.2). Thus in a sequence of Bernoulli trials we may either fix n and take r as being a random variable (R) or fix r and take n as being a random variable (N). This illustrates the need in all modelling to specify very clearly which quantities are random and which are deterministic and fixed.

Before leaving this process let us return briefly to the idea of likelihood. Suppose we have a set of independent observations $n_1, n_2, \ldots n_k$ from a

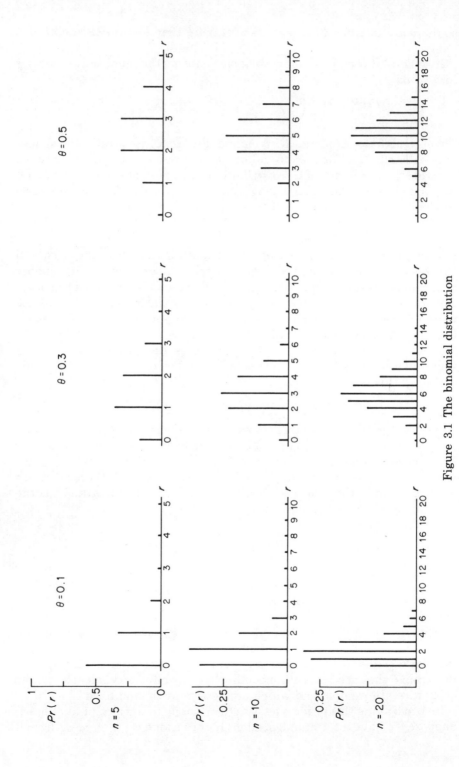

Figure 3.1 The binomial distribution

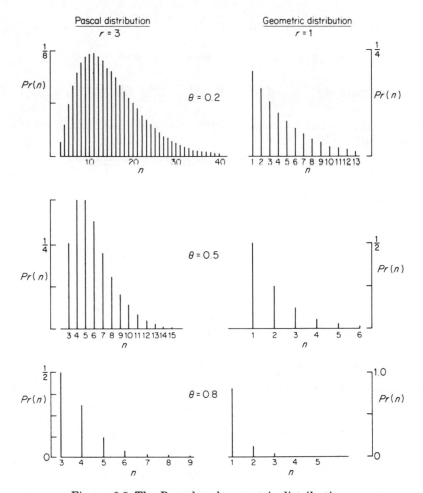

Figure 3.2 The Pascal and geometric distributions

geometric model, the probability and hence the initial likelihood, is

$$\mathscr{L}^*_G(\theta) = (1-\theta)^{n_1-1}\theta \cdot (1-\theta)^{n_2-1}\theta \ldots (1-\theta)^{n_k-1}\theta$$
$$= (1-\theta)^{m-k}\theta^k$$

where $m = n_1 + n_2 + \ldots + n_k$ is the total number of trials. We can thus write

$$L(\theta) = (1-\theta)^{\text{(number of failures)}}\theta^{\text{(number of successes)}}.$$

Similarly given a set of observations on Pascal distributions $n_1, n_2 \ldots, n_k$ when the set values of r were $r_1, r_2 \ldots r_k$ we have the probability, and hence initial likelihood, as

$$\mathscr{L}_{\mathrm{P}}^*(\theta) = \binom{n_1-1}{r_1-1}\theta^{r_1}(1-\theta)^{n_1-r_1} \cdots \binom{n_k-1}{r_k-1}\theta^{r_k}(1-\theta)^{n_k-r_k}$$

$$= \text{constant} \cdot \theta^{\,(\text{number of successes})}(1-\theta)^{\,(\text{number of failures})}$$

where constant $= \binom{n_1-1}{r_1-1}\binom{n_2-1}{r_2-1} \cdots \binom{n_k-1}{r_k-1}$

We now turn to the Binomial Distribution and imagine a sequence of experiments with fixed $n_1, n_2, \ldots, n_k$ giving data (n_1, r_1), $(n_2, r_2), \ldots, (n_k, r_k)$. We may write the probability and hence the initial likelihood as

$$\mathscr{L}_{\mathrm{B}}^*(\theta) = \binom{n_1}{r_1}\theta^{r_1}(1-\theta)^{n_1-r_1} \cdot \binom{n_2}{r_2}\theta^{r_2}(1-\theta)^{n_2-r_2} \cdots \cdots$$

$$\binom{n_k}{r_k}\theta^{r_k}(1-\theta)^{n_k-r_k},$$

which may be rewritten as

$$\mathscr{L}_{\mathrm{B}}^*(\theta) = \text{constant} \cdot \theta^{\,(\text{number of successes})}(1-\theta)^{\,(\text{number of failures})},$$

where constant $= \binom{n_1}{r_1} \cdot \binom{n_2}{r_2} \cdots \binom{n_k}{r_k}$.

We thus have a situation where the probabilities are different for the three distributions but the form, seen as a function of θ, is common. The constants are irrelevant if our interst is in comparing the value of $\mathscr{L}(\theta)$ for different θ. Most uses of the likelihood function are based on such comparisons and hence we can usually ignore such multiplying quantities which do not involve the parameter of interest. Hence for all Bernoulli situations

$$\mathscr{L}(\theta) \propto \theta^{\,\text{number of successes}}(1-\theta)^{\,\text{number of failures}}.$$

Thus the likelihood, and hence any conclusions that we reach from the likelihood, depends only on the numbers of successes and failures in the Bernoulli Process and not on the form of experiment, leading to the Geometric, Pascal or Binomial Distributions.

3.3 The Poisson process and related distributions

The Bernoulli Process considered a series of trials that we could represent as taking place at points $x = 1, x = 2, \ldots$, along an axis. For example we might have a time axis with hourly trials at each of which an event S may or may not occur. Suppose we now consider a continuous axis on which events, S, may occur at any point. For example the axis may represent time and the events could be the occurrence of faults on a production line, or the arrival of telephone calls at a switchboard or of α-particles at a particle counter. As with the Bernoulli Process we will assume that the events occur independently of each other. It is also, similarly, assumed that the rate of occurrence of the events remains constant, at λ per unit time. This continuous analogue of the Bernoulli Process is called a *Poisson Process*. We obtained a Binomial Distribution from the Bernoulli Process by counting the number of events in a fixed number of trials. Similarly the *Poisson Distribution* is obtained by counting the number of events, r, in a fixed time t. The mean number of events in t is λt and the distribution of r is given by

$$Pr(R = r) = e^{-\lambda t}\frac{(\lambda t)^r}{r!} \qquad r = 0, 1, 2, \ldots.$$

Notice that there is no upper limit to the values taken by r. The derivation of the formula for the Poisson Distribution from the Poisson Process is given in many statistical texts, e.g. Parzen.

Another question arises about the observations in the interval of t and that concerns their positioning within the interval. It follows from the assumptions of independence and constant rate that each observation has equal chance of being anywhere in the interval. Thus if we let X be the coordinate of an observation its probability density function will take the form of a horizontal line between 0 and t. To get the total probability, area, equal to one we must therefore have

$$f(x) = \begin{array}{ll} \dfrac{1}{t} & 0 \leqslant x \leqslant t \\ 0 & \text{elsewhere} \end{array}$$

This distribution is called the *Uniform* or *Rectangular Distribution*. The general Uniform Distribution will be defined as

$$f(x) = \begin{array}{ll} \dfrac{1}{\beta - \alpha} & \alpha \leqslant x \leqslant \beta \\ 0 & \text{otherwise} \end{array}$$

and is illustrated in Figure 3.3. We symbolize this by $x \cap \mathrm{U}(\alpha, \beta)$

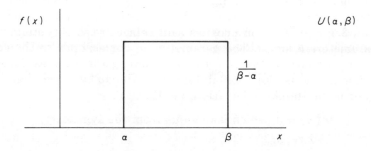

Figure 3.3 The uniform of rectangular distribution

Instead of fixing time and counting the number of events we may fix the number of events and take as the random variable the time, T, for these events to occur. Consider, as the simplest case, the time from the start of the process to the first event. It can be shown that this distribution is given by

$$f(t) = \lambda e^{-\lambda t}, \quad t \geqslant 0.$$

It is called the *Exponential Distribution*, symbolized by $x \sim \text{Expn}(\lambda)$, Figure 2.3 illustrated the shape of the distribution. It is seen that the Exponential Distribution is the continuous time analogy of the Geometric Distribution of the last section.

Consider now the likelihoods arising from a Poisson Process. We discussed the likelihood of the Exponential Distribution in section 2.6 and plotted it in Figure 2.8. The general form for the Exponential Distribution is

$$\mathscr{L}(\lambda) = \lambda^n e^{-\lambda \Sigma t_i}, \qquad \lambda \geqslant 0$$

assuming a set of independent observations $t_1, t_2, \ldots t_n$. For the Poisson distribution with independent observations $r_1, r_2, \ldots, r_n$ the probability and hence the initial likelihood is

$$\mathscr{L}^*_{(\mu)} = e^{-\mu} \frac{\mu^{r_1}}{r_1!} \cdot e^{-\mu} \frac{\mu^{r_2}}{r_2!} \cdots e^{-\mu} \frac{\mu^{r_n}}{r_n!}.$$

As argued, in section 2.6 factors such as $r_1!, \ldots, r_n!$ which do not depend on the parameter, μ, can be ignored when examining the shape of the likelihood function, so we write:

$$\mathscr{L}(\mu) = e^{-n\mu} \mu^{\Sigma r_i}, \qquad \mu \geqslant 0.$$

Figure 3.4 shows the shape of this likelihood function. The maximum point occurs at $\mu = \Sigma r / n$ but the shape naturally depends on n, the larger n the smaller the relative likelihood of values of μ away from this maximum.

3.4 The normal distribution

The Normal Distribution, as its name implies, is the commonest distribution in use. It is a continuous distribution with a symmetrical bell shape, illustrated in Figure 3.5. The position and shape are solely determined by two parameters μ and σ. The parameter, μ, corresponds to the central position of the distribution and the parameter σ, determines the spread about that middle as illustrated in Figure 3.3. It can be shown that μ is the mean and σ the standard deviation, i.e. that

$E(X) = \mu$, which is obvious from the symmetry,

$V(X) = \sigma^2$.

We shall denote such a normal distribution by $N(\mu, \sigma^2)$. If we define a new

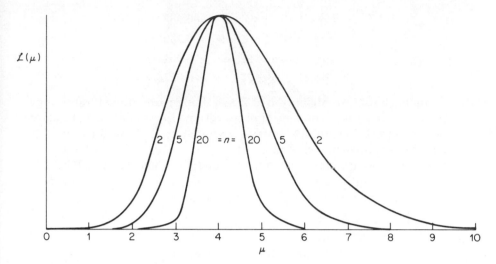

Figure 3.4 The form of the likelihood of the Poisson distribution for varying n

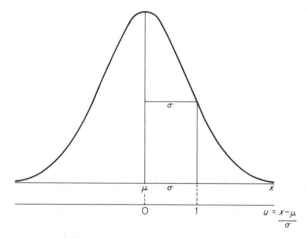

Figure 3.5 The normal distribution

variable, u, with origin at the central point and a scale chosen so that on
Figure 3.3 the point $x = \sigma$ is now one unit out then

$$u = \frac{x - \mu}{\sigma},$$

It can be shown that, and it is intuitively reasonable from the figure that,
the variable u has a distribution $N(0, 1)$. This is called the standard normal
distribution. From tables of the standard normal, given in all books of
statistical tables, it is worth observing that

$$Pr(-1 \leqslant u \leqslant 1) = 0.683$$
$$Pr(-1.96 \leqslant u \leqslant 1.96) = 0.950$$
$$Pr(-2 \leqslant u \leqslant 2) = 0.954$$
$$Pr(-3 \leqslant u \leqslant 3) = 0.9974$$

Referring back to the distribution of X and using percentage terminology we could say that just over 30% of observations lie outside one standard deviation from the mean, 5% outside two standard deviations and $\frac{1}{4}$% outside three standard deviations.

Table 3.1 gives the formulae for the distributions discussed in this chapter and Table 3.2 gives their expectation, variance and skewness. Formulae for some further distributions in common use are given in Tables 3.3 and 3.4.

3.5 Empirical distributions

The distributions that have been described so far arise from situations with a clearly defined probabilistic structure, such as the Poisson Process. They are thus the product of a conceptual modelling approach. We shall later discuss how such models can be identified as suitable for sets of data and then fitted to such data. This identification and fitting is essentially an

Table 3.1 The common distributions of Chapter 3

Distribution	Symbol	Random variable	Range	p.d.f.	Conditions
Binomial	$\mathrm{Bin}(n, \theta)$	R	$r = 0, 1, .., n$	$\binom{n}{r}\theta^r(1-\theta)^{n-r}$	$0 \leqslant \theta \leqslant 1$
Geometric	$G(\theta)$	N	$n = 1, 2, \ldots$	$(1-\theta)^{n-1}\theta$	$0 \leqslant \theta \leqslant 1$
Pascal	$\mathrm{Pas}(r, \theta)$	N	$n = r, r+1, ..$	$\binom{n-1}{r-1}\theta^r(1-\theta)^{n-r}$	$0 \leqslant \theta \leqslant 1$
Poisson	$P(\mu)$	R	$r = 0, 1, 2, ..$	$e^{-\mu}\dfrac{\mu^r}{r!}$	$\mu \geqslant 0$
Uniform (rectangular)	$U(\alpha, \beta)$	X	$\alpha \leqslant x \leqslant \beta$	$\dfrac{1}{\beta - \alpha}$	—
Exponential	$\mathrm{Expn}(\lambda)$	X	$0 \leqslant x \leqslant +\infty$	$\lambda e^{-\lambda x}$	$\lambda \geqslant 0$
Normal	$N(\mu, \sigma^2)$	X	$-\infty \leqslant x \leqslant +\infty$	$\dfrac{1}{\sqrt{2\pi}\sigma} \times \exp -\dfrac{1}{2}\left(\dfrac{x-\mu}{\sigma}\right)^2$	$\sigma > 0$

Table 3.2 Summarizing values for common distributions

Distribution	Expectation	Variance	Coefficient of skewness
Binomial Bin(n, θ), R	$n\theta$	$n\theta(1-\theta)$	$\dfrac{1-2\theta}{\sqrt{n\theta(1-\theta)}}$
Geometric G(θ), N	$\dfrac{1}{\theta}$	$(1-\theta)/\theta^2$	$\dfrac{2-\theta}{\sqrt{(1-\theta)}}$
Pascal Pas(r, θ), N	r/θ	$r(1-\theta)/\theta^2$	*
Poisson P(μ), R	μ	μ	$1/\sqrt{\mu}$
Uniform U(α, β), X	$\dfrac{\alpha+\beta}{2}$	$\beta^2/12$	0
Exponential Expn(λ), X	$1/\lambda$	$1/\lambda^2$	2
Normal $N(\mu, \sigma^2)$, X	μ	σ^2	0

* Complex expression.

Table 3.3 Further distribution models

Distribution	Random variable	Range	p.d.f.	Conditions
Beta	X	$0 \leqslant x \leqslant 1$	$\dfrac{1}{B(\alpha, \beta)}\, x^{\alpha-1}(1-x)^{\beta-1}$	$\alpha, \beta > 0$
Gamma (one-parameter)	X	$0 \leqslant x \leqslant \infty$	$\dfrac{1}{\Gamma(\alpha)}\, x^{\alpha-1}e^{-x}$	$\alpha > 0$
Lognormal	X	$0 \leqslant x \leqslant \infty$	$\log x \cap N(\mu, \sigma^2)$	$\sigma > 0$
Pareto	X	$1 \leqslant x \leqslant \infty$	$\alpha x^{-(\alpha-1)}$	$\alpha > 0$
Gamma (two-parameter)	X	$0 \leqslant x \leqslant \infty$	$\dfrac{\lambda(\lambda x)^{\alpha-1}}{\Gamma(\alpha)}\, e^{-\lambda x}$	$\alpha > 0$ $\lambda > 0$
Negative binomial	R	$R = 0, 1, \ldots$	$\dbinom{r+s-1}{r}\theta^r(1-\theta)^s$	$0 \leqslant \theta \leqslant 1$ $s = 1, 2, \ldots$

empirical process, but when it is complete we shall often consider the fitted model as being informative about the underlying probabilistic structure of the situation. Situations do occur, however, when we have no insight into the mechanisms of probability that might have led to a set of data. Our objective is simply to find a distributional model that matches the shape of the data as accurately as possible. Such distributional models are derived

Table 3.4 Summarizing values for further distribution

Distribution	Expectation	Variance	Coefficient of skewness
Beta	$\alpha/(\alpha+\beta)$	$\alpha\beta/(\alpha+\beta)^2(\alpha+\beta+1)$	$\dfrac{2(\beta-\alpha)(\alpha+\beta+1)^{\frac{1}{2}}}{(\alpha+\beta+2)(\alpha\beta)^{\frac{1}{2}}}$
Gamma (one-parameter)	α	α	$2\alpha^{-\frac{1}{2}}$
Lognormal $w=\exp(\sigma^2)$	$\exp(\mu+\frac{1}{2}\sigma^2)$	$w(w-1)\exp(2\mu)$	$(w+2)(w-1)^{\frac{1}{2}}$
Pareto	$\alpha/(\alpha-1), \alpha>1$	$\alpha/(\alpha-2)-[\alpha/(\alpha-1)]^2$	*
Gama (two-parameter)	α/λ	α/λ^2	$2\alpha^{-\frac{1}{2}}$
Negative binomial	$s\theta/(1-\theta)^2$	$s\theta/(1-\theta)^2$	$(1+\theta)(s\theta)^{-\frac{1}{2}}$

* Complex expression.

on empirical grounds and used for empirical purposes and hence are referred to as *Empirical Distributional Models*. A requirement of such a model is that by changing the parameters a wide range of distributional shapes can be obtained, both in terms of limits on the range of the distribution and in terms of position, spread, skewness and other properties. Various families of empirical distributions have been suggested. Many of these models involve mathematical complexity that put them beyond the scope of this book. However, we shall briefly refer to three such distributions and give references to detailed treatments.

(a) *The Pearson Distributions*

In the early years of this century Karl Pearson suggested that a wide range of distributions could be obtained as solutions to equations of the form

$$\frac{dy}{dx} = g(x, y),$$

where $g(x, y)$ is some suitable function.

For example

$$\frac{dy}{dx} = \left(\frac{\mu-x}{\sigma^2}\right)y$$

leads to the Normal distribution function when the equation is solved for y.

(b) *The Generalized Lambda Family*

It often occurs with the more conceptually based distributions that the p.d.f., $f(x)$ can be simply expressed, but the c.d.f., $F(x)$, and related quantities, which are needed for calculations of probability, are very complicated and require numerical calculation and consequent sets of tables. The Normal distribution is a good example, for though $f(x)$ is relatively simple, $F(x)$ does not have a useful explicit form and numerical approximations have to be used. In practice we often need to use such tables in reverse, finding x for a given $F(x)$. If we require the value, x_p, of the random variable corresponding to a given cumulative probability, p, then

$$F(x_p) = p$$

has to be solved giving

$$x_p = R(p).$$

The value x_p is called the *p-fractile* or, if we use percentages, the *p-percentile*. In the empirical use of distributions the values of x_p for a given p are commonly needed quantities. This suggests that instead of defining a simple distributional form for $f(x)$ we define an empirical distribution via a simple general form for $R(p)$. It is reasonable to make $R(p)$ a function of both p and $1-p$. The Generalized Lambda Family is thus defined by

$$R(p) = \lambda_1 + \{p^{\lambda_3} - (1-p)^{\lambda_4}\}/\lambda_2.$$

The p.d.f. of x does not have a simple explicit form nor does $F(x)$. There are however fairly straightforward methods for estimating the parameters $\lambda_1, \ldots, \lambda_4$, and for handling this model. The approach here, and with other empirical methods, is to introduce the relatively simple mathematical form at the point where it will prove most useful.

(c) *The Johnson System*
The commonality of the Normal distribution, and the vast range of techniques associated with it, suggests that any family of distributions that relate to it in some functional way might be very useful. The Johnson system of distributions relate the random variable of interest, X, to a standard normal variable, Z, by the relation

$$z = \gamma + \eta\, k(x; \varepsilon, \lambda)$$

where $\gamma, \eta, \varepsilon, \lambda$ are the four parameters of the family and k may be one of three relatively simple functional forms, e.g.

$$k_1(x; \varepsilon, \lambda) = \ln\left(\frac{x-\varepsilon}{\lambda}\right)$$

The general shape of the data is used to choose which of the three submodels is appropriate and relatively routine methods are available to estimate the four parameters.

For a detailed study and source references of these methods the reader is referred to Kendall and Steuart for (a), to Shapiro and Gross and to Hahn and Shapiro for (b) and (c).

Chapter 4

Sampling distributions and significance tests

4.1 Distributions arising from normal data

In the previous chapter we examined a number of distributions of single random variables. A wide range of modelling activities involve the collection of data and the calculation of various qualities from that data. We thus start with a random sample $S_1 = (x_1, x_2, \ldots, x_n)$ and from it calculate a quantity t_1. Such a derived quantity is often referred to as a *Statistic*. If we repeated our data collection we would obtain a new sample S_2 and hence a new value for the statistic, t_2. For any given distribution of X and any statistic t we can visualize an infinity of such statistics derived from the infinity of possible samples. There is thus a population of values of t with some probability distribution, $g(t)$. This distribution of the derived random variable is called a *Sampling Distribution*.

Example 1

Consider the situation where we have a sample of n independent observations on a normal population, $N(\mu, \sigma^2)$. Denote the mean value of these observations, the sample mean, by $\bar{x}_1$. If we take a series of samples we could obtain a series of such means $\bar{x}_1, \bar{x}_2, \bar{x}_3 \ldots$. A reasonable question will now be 'what is the distribution of such a set of sample means?' The answer, which we will not prove, is that they are Normally distributed with mean μ and variance $\dfrac{\sigma^2}{n}$. We can regard the $\bar{x}$ as observed values of a random variable $\overline{X}$ and hence state that the sampling distribution of $\overline{X}$ is:

$$\overline{X} \cap N\left(\mu, \frac{\sigma^2}{n}\right).$$

In this particular case it is seen that $\overline{X}$ has the same form of distribution and the same mean as X but that the variance is reduced by a factor $1/n$. It

46

is this factor that is the fundamental reason why we so often use average values, for with the smaller variance we will expect $\bar{x}$ usually to be closer to the population mean μ than any single observation x. The population standard deviation of a sample statistic, such as $\sigma/\sqrt{n}$, is called its population *Standard Error*. From this standard error we see that if we have four observations we would expect 95 % of such calculated averages to be within about $2\dfrac{\sigma}{\sqrt{4}} = \sigma$ of the mean μ, but only about 68 % of individual observations will be within this range. Thus using an average to estimate μ gives a smaller standard deviation. Notice here a 'law of diminishing returns' that operates in almost all the topics that we shall study. If we wish to halve the standard error we have to quadruple n. To reduce it by a third n has to become 9n. Clearly though precision is desirable, the practical costs of data collection frequently put a limit on the precision that we can achive.

The derivation of sampling distributions is a major topic in mathematical statistics and we shall not seek to prove relevant results in this chapter. Our aim here is to introduce some of the most common sampling distributions, most of which relate to sampling from a Normal Population. A major use of sampling distributions is in the significance testing of hypotheses. We shall be referring to such testing at many places in this book so we include in this chapter a very brief discussion of the idea of a hypothesis and of the methodology of significance testing. Almost all introductory texts on statistics deal with this subject at length so we will keep to a brief survey of the topic.

We started the previous example with a statement about independent observations from a Normal distribution. In practice this independence requires that we obtain our samples by a process of random sampling. Thus suppose we sample batches from a production run of a chemical plant, as part of a quality control scheme. If we (i) take the first 10 batches or (ii) take every fiftieth batch or (iii) keep sampling until we have two batches that show poor quality, then we will not have a random sample from the population of all potential batches. We will thus not legitimately be able to use statistics that assume independent observations on the given population. To obtain random samples may require very careful thought about the sampling process and possibly the use of random number tables to decide which items to select. The assumption that independent observations are obtained by random sampling underlies most common statistical techniques. In particular it underlies the use of the common sampling distributions.

We now turn to some further examples of sampling distributions:

Example 2
In discussing the Normal distribution we made the simple transformation

$u = \dfrac{x - \mu}{\sigma}$ to obtain a statistic with a distribution N(0 1). If we apply this to $\overline{X}$ we obtain

$$\overline{u} = \frac{\overline{X} - \mu}{\sigma/\sqrt{n}} \cap N(0, 1).$$

Thus $\overline{X}$ can be standardized by subtracting the mean and dividing by the standard error.

Example 3
Next to the average, $\overline{x}$, the most commonly used sample statistic is that defined as

$$s^2 = \frac{1}{n-1} \sum_{i=1}^{n} (x_i - \overline{x})^2$$

This is used as a measure of data variability and as an estimator of the parameter σ^2 in the Normal distribution. The sampling distribution of s^2 depends on $n - 1$, referred to as the degrees of freedom, and on σ^2. These are related to a distribution called the *Chi-squared Distribution*, written as χ^2_{n-1}, by the statement that

$$(n - 1)s^2/\sigma^2 \cap \chi^2_{n-1}$$

This is a positively skewed distribution with mean $n - 1$.

Example 4
It frequently occurs that we would like to use the distribution for $\overline{u}$ of Example 2 but we do not know σ^2. A natural procedure is to replace it by its estimator, s^2, referred to in the previous example. Thus the true standard deviation of $\overline{x}$, $\sigma/\sqrt{n}$ is replaced by $s/\sqrt{n}$, referred to as the *Sample Standard Error of* $\overline{x}$. Thus we have a new sample statistic, t_{n-1}, where

$$t = \frac{\overline{x} - \mu}{s/\sqrt{n}}$$

If n is large, say 50-plus, s will be very close to σ and will vary little from sample to sample. In this case it is intuitively clear that the sampling distibution to t_{n-1} will be very close to a N(0, 1). However, as n decreases the sampling distribution of t_{n-1} will develop longer, but symmetrical, tails as the variability of s from sample to sample is more pronounced. The distribution of t_{n-1} thus varies with n. The determining factor is expressed as $n - 1$ which is called the *degrees of freedom*. The distribution of t_{n-1} is called the *t-distribution*.

Example 5
A further sampling distribution that we shall need to make use of, is the *F*-distribution. This arises when comparing variances of normal distri-

butions. If $s_1{}^2$ is obtained from n observations on a $N(\mu_1, \sigma^2)$ and $s_2{}^2$ independently from m observations on a $N(\mu_2, \sigma^2)$ with the same σ^2 then

$$\frac{s_1{}^2}{s_2{}^2} \cap F_{n-1, m-1}$$

It will be seen that the distribution of F depends on the two degrees of freedom of $s_1{}^2$ and $s_2{}^2$, but does not depend on μ_1, μ_2 or σ^2.

The distributions of y, t, χ^2 and F are included in standard sets of statistical tables. The mathematical formulae for these distributions are included in most texts on the theory of statistics. For most modelling purposes these mathematical forms are not needed, what is required is an appreciation of the general shape of the distributions and the use of their tables. The general shape of these distributions, together with many other useful distributions is given in Hastings and Peacock.

All the examples considered give the exact distributions of sample statistics. Often sample statistics have very complex distributions and it is a convenient fact that many of these approximate to a Normal distribution for statistics calculated from large samples.

Example 6
We will later (section 10.4) refer to a measure of relationship called the sample correlation coefficient, r, derived from paired observations (x_1, y_1) (x_2, y_2), . . . , (x_n, y_n). The distribution of this is highly involved. However,

for large n and independent data r is approximately $N\left(0, \dfrac{1}{\sqrt{n-1}}\right)$.

4.2 Hypotheses

If we take a purely empirical approach to modelling we shall simply look at the data without any prior assumptions. If, however, we take a conceptual approach, we approach our data with certain theories about how it will behave. We shall thus look to the data to support or possibly falsify our theories. The more precisely we can formulate our ideas and theories the more effectively shall we be able to test them against the evidence of the data. Such a precise means of formulation is provided by the idea of a statistical hypothesis. The nature of such hypotheses is best indicated by some examples.

Examples
Past experience indicates that the concentration of a salt in solution produced by a particular chemical plant is normally distributed with a known mean μ_0 and variance σ_0^2. Given twenty samples, with concentrations, x_i, produced by a different plant we may wish to question whether the model, $N(\mu_0, \sigma_0^2)$, is suitable for the new data. We may thus formulate a

sequence of hypotheses about the random variable X, representing output from the second plant

H_{01}: X is Normally distributed.

H_{02}: $V(X) = \sigma_0^2$, assuming Normality.

H_{03}: $E(X) = \mu_0$, assuming Normality and variance σ_0^2.

or

H'_{03}: $E(X) = \mu_0$, assuming Normality but not σ_0^2.

If we lacked certainty about μ_0 and σ_0 we might take a set of observations, y_i, from the first plant, using Y to represent the concentrations from that plant. With the two random variables we could formulate an analogous sequence of hypotheses.

H_{04}: X and Y have the same distribution.

or

H'_{04}: Y is Normally distributed.

H_{05}: $V(X) = V(Y)$, assuming both X and Y Normal.

H_{06}: $E(X) = E(Y)$, assuming both Normal with variance σ_0^2.

or

H'_{06}: $E(X) = E(Y)$, assuming both Normal with the same, but unknown, variance.

The previous examples illustrate a number of classifications and features of hypotheses. A hypothesis focusses on certain specific parameters or features of the situation, but it also makes other assumptions about the situation, e.g. the normality assumed in H_{02}. Though the methods we shall develop will examine the hypotheses, they do not examine these assumptions. Such assumptions must therefore be based on very clear evidence or be tested as preliminary hypotheses, as H_{05} precedes H_{06}. Some hypotheses give a precise description of the probability distribution of the observations. Examples H_{02} and H_{03} give such hypotheses, which are termed '*Simple*' *Hypotheses*. In other cases the hypotheses do not give such precision. Thus in Example H'_{03}, though μ was defined, σ^2 was left undefined. In Example H_{04} the actual distributions of X and Y were left undefined. Such non-simple hypotheses are termed *Composite Hypotheses*.

In each of the examples there were implicitly, two hypotheses:

(a) A basic hypothesis, usually simple. This we refer to as the *Null Hypothesis*, H_0. This hypothesis describes what we hold as a tentative belief until such time as the evidence of the data leads us to reject it.

(b) An *Alternative Hypothesis*, usually composite, which is accepted if we reject the null hypothesis. Thus in Example H_{03}, if we reject H_{03}: $E(X) = \mu_0$, then we will be led to accept H_{13}: $E(X) \neq \mu_0$. In example H_{05}, the alternative that would most interest us would be H_{15}: $V(X) \leqslant V(Y)$.

We would refer to $H_{13}: E(X) \neq \mu_0$, as a *Two-sided Alternative* and to $H'_{13}: E(X) < \mu_0$ and to H_{15} as *One-sided Alternatives*.

4.3 Significance tests

We shall now bring together the ideas of the previous sections to develop a methodology for testing hypotheses. The essential idea is illustrated in Figure 4.1 and depends on the construction of a statistic, T, called a *test statistic*, whose sampling distribution is completely known if some hypothesis H_0 is assumed true. We can then say that values in the central portion of this distribution, A, are the values that we expect. Values out in the extreme tails, C, are unlikely to be observed very often. Using our assumptions, our Hypothesis and our data we find the observed numerical value of the statistic, T_{obs}. If T_{obs} lies in A we shall have observed what we

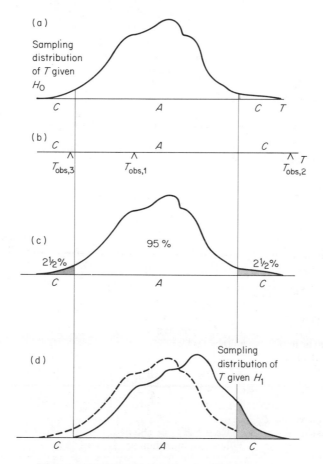

Figure 4.1 The Ideas of a significance test

Table 4.1 Significance test examples

Data	Concentration of a salt in solution (grams/litre). 20 samples								
10.4	9.3	11.2	10.1	10.8	9.7	9.7	8.2	10.2	10.6
9.5	10.0	10.1	10.2	10.3	9.2	10.4	8.8	11.0	8.7

Underlying assumption: X is Normally distributed

(a) Hypothesis H_{02} : $V(X) = \sigma_0^2 = 1.1^2$.
Alternative $\sigma_0^2 \neq 1.1^2$

Test statistic T : $(n-1)s^2/\sigma_0^2$

Distribution H_{02} : χ_{n-1}^2

5% critical region :

$$C \quad 8.9 \qquad\qquad A \qquad\qquad 32.8 \quad C$$

Observed value T_{obs} : 9.7

Conclusion : Accept H_{02}

(b) Hypothesis H_{03} : $E(X) = \mu_0 = 10$,
assuming $\sigma_0 = 1.1$
Alternative $\mu_0 \neq 10$

Test statistic T : $(\bar{x} - \mu_0)/(\sigma_0/\sqrt{n})$

Distribution H_{03} : $N(0, 1)$

5% critical region :

$$C \quad -1.96 \qquad\qquad A \qquad\qquad 1.96 \quad C$$

Observed value T_{obs} : -0.33

Conclusion : Accept H_{03}

(c) Hypothesis H_{03}' : $E(X) = \mu_0 = 10$.
Alternative $\mu_0 \neq 10$

Test statistic T : $(\bar{x} - \mu_0)/(s/\sqrt{n})$

Distribution H_{03}' : t_{19}

5% critical region :

$$C \quad -2.09 \qquad\qquad A \qquad\qquad 2.09 \quad C$$
$$0$$

Observed value T_{obs} : -0.45

Conclusion : Accept H_{03}'

(d) Hypothesis H_{02}^* : $V(X) = \sigma_0^2 = 1.1^2$.
Alternative $\sigma_0^2 < 1.1^2$

Test statistic T : $(n-1)s^2/\sigma^2$

Distribution H_{02}^* : χ_{n-1}^2

5% critical region :

$$C \qquad 10.1 \qquad\qquad A$$

Observed value T_{obs} : 9.7

Conclusion : Reject H_{02}^* at the 5%
level of significance

expected on the basis of our hypothesis, H_0. If, however, T_{obs} lies in the tail regions, C, then we might conclude either (i) that H_0 is true but we have observed a rare event, or (ii) that the observations, as reflected in T_{obs}, are unlikely, given the hypothesis, and hence that hypothesis should be rejected. The procedure of significance testing is to take the latter approach, though we must remember that the former statement may be true. Looking at Figure 4.1(b) it is clear that if the observed value is $T_{obs, 1}$ then we shall continue to hold our hypothesis. If the observed value is $T_{obs, 2}$ then we would reject H_0 with some confidence, though it must always be remembered that the very rare event does still happen. Indeed the more tests carried out the more likely it is to happen. A problem becomes apparent if the observed value is $T_{obs, 3}$. This raises the problem of where to make the division between A and C. There is no natural place to put it. Wherever it goes it will be purely arbitrary. The standard convention, illustrated in Figure 4.1(c), is to choose the boundaries, *critical values*, so that, if H_0 is true, T_{obs} has a small clearly defined probability, say 0.05, of being in C. If T_{obs} does then lie in C we say we reject the hypothesis at the 5% *level of significance*. The region C of the T axis is called the *Critical Region* and A the *Acceptance Region*. Critical values for common problems are given in books of statistical tables, e.g. Neave. Notice that rejecting the hypothesis at the 5% level does not imply that H_0 is false. The 5% tells us that if H_0 is true we will incorrectly reject H_0 on 5% of occasions. Conversely if we accept H_0, this does not prove that H_0 is true, indeed it is seen from Figure 4.1(d) that, if the true situation was given by the distribution under some similar hypothesis H_1, we would tend to accept H_0 nearly as frequently as if H_0 were true. The merit of this procedure is not that it proves our hypothesis true or false—no procedure can guarantee that—but rather that it provides a conventional means of communicating and acting on the consistency or inconsistency of our data with our tentative hypotheses.

Let us now illustrate the above ideas by returning to our examples. Table 4.1 shows four tests of hypotheses relating to a single set of data on concentrations. Past knowledge of the production process specifies a $N(10, 1.1^2)$ model. A modification to the process leads to the sample on which tests are carried out to see whether it is consistent with the original model. Notice the difference between examples (a) and (d). In (a) the alternative is two-sided, in (d) it is one-sided. Even though the data and null hypothesis remain the same the change in the alternative leads to a change in the critical region and hence in the conclusion. The use of a one-sided alternative is sensible in this situation since interest would naturally lie in whether the variability of the sample was less than the standard represented by $\sigma_0 = 1.1$.

Chapter 5

Model composition using random variables

5.1 Introduction

When we build a toy model from a kit, the kit will contain a variety of components and also such things as glue, clips, nuts, and bolts that enable the components to be fastened together to compose the final model. The same components may be used to make a variety of models, some rather ridiculous, by putting them together in many different ways. Different means of joining components lead to different properties. Rivets fix firmly, hinges allow a rotation of parts. With this analogy in mind we look at the ways in which the components of the previous sections may be combined in a model. We consider in particular simple addition and multiplication. This may seem rather trivial but we shall see, as we progress, that many fundamental properties and useful practical aspects of models relate directly to the composition of the model in terms of how the components are joined together.

5.2 Additivity

The simplest and commonest method of relating model components is to add them together. This method superimposes the effects of the various components but they still act essentially independently. For example if $x = u + v$ an increase in u of two units will increase x by two units, irrespective of the value of v, this is not true for say a multiplicative model $x = uv$.

Example 1
Consider the data in Table 5.1. This gives the results of an experiment in which a chemical property is measured for the sixteen combinations of four chemical treatments in the presence of four catalysts. The simplest model for this experiment is based on the assumption that the treatment and the

54

Table 5.1 A model of experimental data

(a) *The data*

Treatment chemical i

x_{ij}		1	2	3	4
	1	52.7	53.0	50.8	51.9
	2	52.6	52.2	50.7	51.6
Catalyst j	3	54.6	52.4	51.9	51.7
	4	51.5	52.5	50.5	49.6

(b) *The model*

Treatment chemical i

$E(x_{ij})$		1	2	3	4
	1	$\mu+\gamma_1+\beta_1$	$\mu+\gamma_2+\beta_1$	$\mu+\gamma_3+\beta_1$	$\mu+\gamma_4+\beta_1$
	2	$\mu+\gamma_1+\beta_2$	$\mu+\gamma_2+\beta_2$	$\mu+\gamma_3+\beta_2$	$\mu+\gamma_4+\beta_2$
Catalyst j	3	$\mu+\gamma_1+\beta_3$	$\mu+\gamma_2+\beta_3$	$\mu+\gamma_3+\beta_3$	$\mu+\gamma_4+\beta_3$
	4	$\mu+\gamma_1+\beta_4$	$\mu+\gamma_2+\beta_4$	$\mu+\gamma_3+\beta_4$	$\mu+\gamma_4+\beta_4$

catalysts have independent additive effects on the measurements. Thus if x_{ij} is the observation obtained applying treatment i and catalyst j, then the model is

$$x_{ij} = \mu+\gamma_i+\beta_j+e_{ij}, \qquad i = 1, 2, 3, 4$$
$$j = 1, 2, 3, 4$$

where μ is the mean value,

γ_i is the effect of treatment i on the mean,

β_j is the effect of catalyst j on the mean,

e_{ij} is the random variation.

To make sense of the definition of μ we require that $\sum_i \gamma_i = \sum_j \beta_j = 0$. These parameters thus indicate the way a treatment or a catalyst change the treatment mean $(\mu+\gamma_i)$ or batch mean $(\mu+\beta_j)$ from the overall mean (μ). It is clear from the additive nature of this model that the effect of the different treatments increases or decreases the catalyst mean, $\mu+\beta_j$, by the same amount for each catalyst. This additivity enables us to disentangle the effects of treatments and catalysts. For example we will show that the difference between the overall average, denoted by $\bar{x}_{..}$, and the average of the observations for the first treatment, denoted by $\bar{x}_{1.}$, provides a simple estimate of γ_1.

Additive models possess a number of properties that are very simple and useful. We will state some of these results, the reader will find proofs in texts on mathematical statistics given in the references. The notation follows our usual practice with X_1, X_2 etc. denoting random variables, α, β etc. here denoting constants.

(a)
$$E(X_1 + X_2) = E(X_1) + E(X_2)$$
and more generally
$$E(\alpha + \beta X_1 + \gamma X_2) = \alpha + \beta E(X_1) + \gamma E(X_2).$$

Thus to find the expected value of an additive model we simply need the expected values of the individual terms.

(b) If X_1 and X_2 are independent then
$$V(X_1 + X_2) = V(X_1) + V(X_2)$$
and more generally
$$V(\alpha + \beta X_1 + \gamma X_2) = \beta^2 V(X_1) + \gamma^2 V(X_2).$$

The constant α disappears on the right-hand side because the variance of a constant is zero, the β^2, γ^2 appear because variance has the dimension of the square of the variable. Thus, as in (a), the variance can be found by adding the variances of the individual terms.

If the condition of independence does not hold then there has to be a modification to allow for the covariance of X_1 and X_2 thus:
$$V(\beta X_1 + \gamma X_2) = \beta^2 V(X_1) + 2\beta\gamma\, C(X_1, X_2) + \gamma^2 V(X_2).$$

(c) If $X_1 \cap N(\mu_1, \sigma_1^2)$, $X_2 \cap N(\mu_2, \sigma_2^2)$ independently then
$$\alpha + \beta X_1 + \gamma X_2 \cap N(\alpha + \beta\mu_1 + \gamma\mu_2, \beta^2\sigma_1^2 + \gamma^2\sigma_2^2).$$

This result adds to those of (a) and (b) the fact that the sum of Normally distributed observations is itself Normal. The Normality still holds when the variance is adjusted as in (b) for the non-independent case.

(d) Clearly the results of (a), (b), (c) can be extended to sums of many variables. Thus if we have n independent random variables, each $N(\mu, \sigma^2)$, then
$$\sum_1^n X_i \cap N(n\mu, n\sigma^2).$$

A very remarkable result, called the *Central Limit Theorem*, shows that, apart from some rare exceptions, the above result is approximately true for large n even when the random variables X_i are not Normally distributed. This result often enables us to use Normal distribution results for models where the components are not themselves normal.

We now illustrate the use of these rules.

Example 2

Suppose $X_i \cap U(0, 1)$ and the values of X_i are the pseudo-random numbers generated on a computer or calculator. For this distribution (Table 3.2)
$$E(X_i) = \tfrac{1}{2}, \qquad V(X_i) = \tfrac{1}{12}.$$

So if we generate 12 such numbers and add we obtain

$$Z = \sum_{1}^{12} X_i, \qquad E(Z) = 6 \qquad V(Z) = 1.$$

By the Central Limit Theorem:

$$Z \cap N(6, 1),$$

approximately. In fact the approximation is quite good and this is the means adopted on some computers to simulate a Normal distribution, that is to say to generate numbers whose distribution is Normal.

Example 3
Consider the sample average obtained from n independent observations all having mean μ and variance σ^2.

$$\bar{X} = \frac{1}{n} \sum_{1}^{n} X_i$$

By (a) $E(\bar{X}) = \frac{1}{n} \sum_{1}^{n} E(X_i)$

$$= \mu.$$

By (b) $V(\bar{X}) = \frac{1}{n^2} \sum_{1}^{n} V(X_i)$

$$= \frac{\sigma^2}{n}.$$

By (c) and (d) $\bar{X} \cap N\left(\mu, \frac{\sigma^2}{n}\right)$, exactly so if the X_i were Normal and approximately so if not. The quality of the approximation improves as the number of observations used for the average increases.

Example 4
Consider again Example 1, the chemical experiment model of Table 5.1. We have

$$E(\bar{X}_{..}) = E\left[\frac{1}{16} \sum_{ij} X_{ij}\right]$$

$$= E\left[\frac{1}{16} \sum_{ij} (\mu + \gamma_i + \beta_j + e_{ij})\right]$$

$$= \mu + \frac{4}{16} \sum_{i} \gamma_i + \frac{4}{16} \sum_{j} \beta_j + \frac{1}{16} \sum_{ij} E(e_{ij})$$

$$= \mu.$$

Similarly for a specific treatment average

$$E(\bar{X}_{i.}) = E\left[\frac{1}{4}\sum_j X_{ij}\right]$$

$$= E\left[\frac{1}{4}\sum_j (\mu + \gamma_i + \beta_j + e_{ij})\right]$$

$$= \mu + \gamma_i + \frac{1}{4}\sum \beta_j + \frac{1}{4}\sum_j E(e_{ij})$$

$$= \mu + \gamma_i$$

Hence $E(\bar{X}_{i.} - \bar{X}_{..}) = \gamma_i$.

Thus we have made use of the various additive properties to show that treatment effects, γ_i, can be isolated by comparing the treatment mean and the overall mean. Care needs taking if we wish to find the variance of the quantity $E(\bar{X}_{i.} - \bar{X}_{..})$ since the two terms contain the same X_{ij} and are thus not independent. To examine this we note that if we treat the terms μ, γ, β as constants, whose variance is zero, we will have, specifying $i = 2$ for clarity,

$$V(\bar{X}_{2.} - \bar{X}_{..}) = V\left[\frac{1}{4}\sum_j e_{2j} - \frac{1}{16}\sum_{ij} e_{ij}\right]$$

$$= V\left[\frac{3}{16}\sum_j e_{2j} - \frac{1}{16}\sum_j (e_{1j} + e_{3j} + e_{4j})\right]$$

$$= \frac{9}{16^2}(4\sigma^2) + \frac{1}{16^2}(12\sigma^2)$$

$$= \frac{3}{16}\sigma^2$$

where $V(e_{ij}) = \sigma^2$.

A number of distributions other than the Normal also have useful additive properties. In particular

(i) The Poisson: if $R_1 \cap P(\mu_1)$, $R_2 \cap P(\mu_2)$ independently

$$R_1 + R_2 \cap P(\mu_1 + \mu_2)$$

Thus Poisson variables add in a similar fashion to normal variables. This result is intuitively clear if one imagines two independent Poisson processes occurring at the same time. The number of events and the rate of occurrence in the combined process will be the sum of those in the individual processes.

(ii) The Binomial: if $R_1 \cap \text{Bin}(n_1, \theta)$, $R_2 \cap \text{Bin}(n_2, \theta)$ independently

$$R_1 + R_2 \cap \text{Bin}(n_1 + n_2, \theta)$$

The intuitive justification for this is seen by imagining two experiments in a Bernoulli situation with n_1 and n_2 trials. This is equivalent to a single experiment of $n_1 + n_2$ trials provided both experiments have the same value for the parameter θ.

(iii) The χ^2 distribution: we have already briefly referred to a distribution called the χ^2 distribution, pronounced 'chi-squared'. This relates to the Normal distribution in the following way. If $u_1, u_2, \ldots, u_n$ are independent standard normal variables $N(0, 1)$ then

$$u_1^2 + u_2^2 + \ldots + u_n^2 \cap \chi_n^2$$

where the single parameter controlling the shape of the χ^2 distribution, the degrees of freedom, takes the value n. As u relates to X_i from a $N(\mu, \sigma^2)$ we have

$$\sum_1^n \left(\frac{X_i - \mu}{\sigma} \right)^2 \cap \chi_n^2$$

It can also be shown that

$$\sum_1^n \left(\frac{X_i - \bar{X}}{\sigma/\sqrt{n}} \right)^2 \cap \chi_{n-1}^2$$

That is to say if we replace the population mean μ, usually unknown, by the sample mean, $\bar{X}$, the distribution is still a χ^2 but the degrees of freedom are reduced to $n - 1$.

The way we defined the χ^2 implies that if we have two independent χ^2 variables χ_m^2 and χ_n^2, with m and n degrees of freedom then

$$\chi_m^2 + \chi_n^2 \cap \chi_{m+n}^2.$$

Thus the χ^2 distribution also has an additivity property.

5.3 Multiplicativity

The major role of additivity in statistical modelling becomes most apparent when one considers alternative forms of model composition. It would be possible to expand the previous section into a whole book. However with multiplicative structures there are a few general results, and some of these apply because of the ability of logarithms to convert a multiplication to an addition. We now look briefly at these results.

(a) $E(X_1 X_2) = E(X_1) E(X_2)$: this applies only if X_1 and X_2 are independent of each other. If they are not independent and have correlation ρ and variances σ_1^2 and σ_2^2, then a little algebra on the definition of ρ will show that

$$E(X_1 X_2) = E(X_1)E(X_2) + \rho \sigma_1 \sigma_2$$

Thus a positive correlation will inflate the value of $E(X_1 X_2)$ above the independent case, and vice versa. Note that this adjustment will be negligible if the standard deviations are much less than the expectation.

The ratio $\dfrac{\sigma}{E(X)}$ is called the *Coefficient of Variation*, v. The percentage error in $E(X_1 X_2)$ introduced by ρ is thus $100\rho\, v_1\, v_2$. Hence 10% coefficients allow at worst a 1% error.

(b) In the same way that the normal distribution is important in additive situations, a distribution called the *Lognormal* is important in multiplicative situations. The simplest way of defining this distribution for a random variable X is to say that X is lognormal if $\log X$ is Normal, $N(\mu, \sigma^2)$. Table 3.4 gives the properties of this distribution. In section 5.2(d) we noted that if enough random variables are added together they lead to a normal distribution. An equivalent result is that if enough random factors influence a situation in a multiplicative fashion, the ultimate random variable will tend to have a lognormal distribution. Such multiplicative effects tend to act in economic situations where changes are often percentage changes, which are essentially multiplicative.

Chapter 6

Regression models

6.1 The linear model

In the first chapter we investigated a simple model whose deterministic part was a straight line, $y = \theta x$, through the origin. The randomness in the situation was represented by a simple added 'error' term, $+e$. Using the terminology that we have now developed it is reasonable, from a conceptual modelling viewpoint, to make four assumptions about e:

A1: The errors $e_1, \ldots, e_n$ associated with any set of data are an independent random sample from the distribution of e. They are thus independent, identically distributed (*iid*).
A2: The random variable e has zero mean, $E(e) = 0$.
A3: The variance of e is constant, $V(e) = \sigma^2$.
A4: The random variable e is not influenced by the regressor variable x, nor vice versa.

Throughout this book the *Error Variable*, e, will be assumed to satisfy these assumptions unless stated otherwise.

The final model is written as

$$y_i = \theta x_i + e_i, \qquad e_i \cap iid\,(0,\,\sigma^2).$$

If the line did not go through the origin we could write

$$y_i = \beta_0 + \beta_1 x_i + e_i$$

where β_0 is the intercept and β_1 the slope. This two-parameter model is often called the *Simple Linear Regression* of the dependent variable y on the regressor variable x. Almost all computers and many pocket calculators have programs for fitting and using this model. Table 6.1 shows a range of different specific equations illustrating its use and extension.

It must be emphasized that the use of the word linear usually refers not to straight line of y on x but to the linear influence of the parameters. The equation is linear in the parameters β_0 and β_1, called *Regression*

61

Table 6.1 Examples of regression models

Dependent variable y	Regressor variables x	Equation
Radioactivity of blood tagged with radioisotope	Number of days from transfusion	$y = 94.0 - 10.2x + e$
Breaking stress of casting	Breaking stress of test piece	$y = 26.52 + 0.432x + e$
% locusts killed	Strength of poison	$y = -23.3 + 9.4x + e$
Density of fluid	Measure of light transmitted	$y = -36.5 + 31.8x - 6.5x^2 + e$
Auction income	Number of items sold in category j, $j = 1, 2, 3$.	$y = 2.4 + 3.2x_1 + 1.0x_2 + 0.8x_3 + e$

Coefficients. It must also be emphasized that the influence of random error in the model is allowed for only in the error term, e. The regressor variables are thus treated as deterministic. To do this we require that one of the following is true:

(a) The regressor variables are deterministic quantities that can be measured with negligible error, e.g. length, time, etc.
(b) The regressor variables are under complete control, e.g. if we are regressing chemical output on reaction temperature x we must assume that x_i is the true temperature and that we can accurately set the temperature of the reaction.
(c) The regressor variables are random variables that (i) can be accurately measured (ii) are seen and will be used as the determinants of the dependent variable y, i.e. we shall always be using the regression model to predict y given known or specified values of x.

It follows from the previous points that the regression equation relates the values of y to a function of some known values which we have called x_i. These known values may however be mathematical functions of some other known values. If they are the equation still satisfies our requirements for the linear regression model. The following equations are hence also simple linear regression models:

$$y_i = \beta_0 + \beta_1 \log x_i + e_i$$
$$y_i = \beta_0 + \beta_1 e^{3x_i} + e_i$$
$$y_i = \beta_0 + \beta_1 / x + e_i$$
$$y_i = \beta_0 + \beta_1 x^2 + e_i$$

The two-parameter model involves only one regressor variable. In many situations the dependent variable y may depend on several variables. Thus each observation on the dependent variable and p regressor variables will take the form $(y_i, x_{i1}, \ldots, x_{ip})$. Assuming that the relation is linear the model will be:

$$y_i = \beta_0 + \beta_1 x_{i1} + \quad + \beta_p x_{ip} + e_i$$

Using the convention that $x_{0i} = 1$, all i the model can conveniently be rewritten using summation notation as

$$y_i = \sum_{j=0}^{p} \beta_j x_{ij} + e_i, \quad e_i \cap iid\,(0, \sigma^2).$$

The linear regression model is probably the most commonly used model. There are four main grounds for this common use:

(a) There are many situations where the linear regression model accurately describes how variables do relate.

(b) There are many situations where the relationship is not truly linear but for which the linear model gives an adequate approximation provided there are constraints on the range over which the regressor variables are allowed to vary. It is important in these situations that this range is clearly specified as part of the statement of the model. Obviously if the model is used outside this range the user may be misled in his conclusions. For example, many of the regressions in Table 6.1 give nonsense values near $x = 0$.

(c) The model provides a simple starting point from which to explore more elaborate models. The clearest way to see small curvature in a line is to draw a straight line next to it. The clearest way to see non-linearities in models is to compare with a fitted linear model.

(d) The linear model is in fact a family of models. Consider the models

$$y = \beta_0 + \beta_1 x_1 + \beta_2 x_2 + \beta_3 x_3 + e$$
$$y = \beta_0 + \beta_1 x_1 + \beta_2 x_2 + e$$
$$y = \beta_0 + \beta_1 x_1 + e$$

it is seen that each model is a special case of the previous one with the final β set at zero. This ability to build up, or down, within the linear regression family of models has great benefits throughout the modelling process.

6.2 The regressor variables

There are many different ways in which regressor variables are selected for a regression model. We will look at the selection of variables in Chapter 12. For the present it is sufficient to discuss the types of variables that can be used in regression models. There are two extreme situations and many that combine both. On the one extreme the regression model represents a direct cause and effect relationship. The amount of run-off water in a drainage basin will have rainfall as a clearly causal regressor variable. The amount of product chemical from a chemical reaction will have the amount of reacting chemical as a causal regressor variable. At the other extreme we may simply observe a linear relationship between two variables in a complex situation. We cannot say that either causes the other, in all probability both are influenced by a number of common factors, that produce common behaviour and hence the linear relationship. A classical example is the use in business of *Lead Indicators* in which certain specific economic variables, such as starts on new building, relate to measures of the state of a whole economy some months later. In these situations choice of which variable is treated as dependent, which as regressor, is a function of the use to which the model is to be put. Thus for forecasting the future of the economy the Lead Indicator is used as the regressor variable. Many situations lie between these two extremes. Our understanding of the prototype suggests a linear relationship between two variables, but there is not a simple cause and effect. There may be many environmental factors, not explicit in the regression, that are influencing both dependent and regressor variables but there is also a direct link. For example we may regress the marks in a major examination on the score in an IQ test. We might expect a relation but could not argue for the IQ to be either the only or the direct causal factor.

It may be observed that the interpretation of the error term depends on the interpretation of the nature of the regression model. For the simple causal model the errors may indeed be the errors due to measurement error, or to minor influences from the environment. For the 'observed regression' the error term acts as the term that picks up all the factors and complications that are not included in the simple linear relationship.

Regressor variables can take a range of special forms worthy of separate consideration. The most common forms are:

(a) Functions of a single variable

A relationship may be non-linear in the regressor variable but involve several parameters in a linear model. Perhaps the simplest example is where y is quadratic in x so the model is

$$y_i = \beta_0 + \beta_1 x_i + \beta_2 x_i^2 + e_i.$$

This generalizes to give the *Polynomial Model* of degree p

$$y_i = \beta_0 + \beta_1 x_i + \beta_2 x_i^2 + \ldots + \beta_p x_i^p + e_i$$

A build-up of terms of similar form occurs in the Fourier or Harmonic model. A simple example of this for weekly data, y_i, with i as the week number, takes the form

$$y_i = \beta_0 + \beta_1 \cos 2\pi i/52 + \beta_2 \sin 2\pi i/52$$
$$+ \beta_3 \cos 4\pi i/52 + \beta_4 \sin 4\pi i/52 + e_i.$$

The model is thus a linear addition of a number of pure oscillations.

(b) Dummy variables

A *Dummy* or *Indicator Variable* enables the model to allow for qualitative as well as quantitative factors. The presence or absence of a qualitative feature for a particular observation y_i is denoted by $x_i = 1$ or $x_i = 0$. If the qualitative feature has several categories then a set of dummy variables may be used. For example, if in a regression of house prices on relevant variables, the building material is classified as Brick, Wood or Other Materials, then two dummy variables, x_1, x_2, will suffice,

$$x_1 = 1 \quad x_2 = 0 \quad \text{denoting Brick}$$
$$x_1 = 0 \quad x_2 = 1 \quad \text{denoting Wood}$$
$$x_1 = 0 \quad x_2 = 0 \quad \text{denoting Other Materials}$$

In this situation the coefficient β_1 of x_1 would denote the price differential between houses built by brick and by non-brick materials.

Dummy variables are also used to allow for particular events when the data is a sequence in time. Thus the effect of a strike could be allowed for by using a variable x_k where

$$x_{ik} = \begin{array}{l} 1 \text{ for observations made during strike} \\ 0 \text{ otherwise} \end{array}$$

The coefficient β_k will thus indicate the effect of the strike on the dependent variable. The effect of a change in legislation, tax, etc, that might influence a dependent variable are analysed by including a dummy variable of the form

$$x_{ik} = \begin{array}{l} 0 \text{ before change} \\ 1 \text{ after change} \end{array}$$

Such variables are sometimes called *Intervention Variables* and the study of such 'before and after' effects is called *Intervention Analysis*.

(c) Totally known factors

It sometimes occurs that not only is the variable x_t known but also its coefficient, β_k, is known. As our concern in the model building stage is to investigate the unknown it is advisable to remove the known effects. This is most easily done by moving the term, $\beta_k x_{ik}$ to the left-hand side and defining a new, adjusted, dependent variable $y' = y - \beta_k x_k$. It may be that though we do not know any regression coefficients, we do know a relation between them. For example we may know that $\beta_h = 3\beta_k$. In such a situation we can define a new regressor variable $z = 3x_h + x_k$. The term $\beta_z z$ equals $3\beta_z x_h + \beta_z x_k$ which ensures that the coefficient of x_h is three times that of x_k.

(d) Proxy variables

In some practical situations measurements of an important variable are either unobtainable, too unreliable, or available too late. In such cases it may be possible to find another related variable that does not have such disadvantages. This variable acts as a proxy for the other. For example, in a study of seat belts and their effects on road deaths, an obviously relevant variable was the total vehicle mileage per year. As there was no way of obtaining this, the total petrol usage per year was used as a proxy variable, since this was readily obtainable and clearly relates to mileage. Whenever proxy variables are used, additional implicit assumptions are introduced into the model. For example, the above proxy clearly assumes that the average petrol consumption per mile remains constant over the period covered by the data. This implies an assumption about the constancy of mix of engine sizes and types. If this assumption cannot be made then a more complex model and further information will be required.

(e) Time and time related regressors

The data for many regression models are sequences of observations made over time, t, e.g. as hourly or yearly observations. It is hardly surprising therefore that models for such situations involve t as a regressor variable as well as a subscript. For example sales of car radios, y_t, might be related to new car registrations, x_t, and also to time, to allow for the increasing numbers of cars having radios. The model might be

$$y_t = \beta_0 + \beta_1 x_t + \beta_2 t + e_t$$

or

$$y_t = \beta_0 + (\beta_1 + \beta_2 t) x_t + e_t,$$

which is still of linear form but allows for a proportional increase over time. The terms on the right-hand side of the model all involve values at the current time t. The model thus describes a situation that can be explained in terms of instantaneous pictures and as such it is essentially static. If y_t depended on previous values of x_t then the situation is more dynamic and we cannot regard the model as a simple regression. We shall consider such dynamic models in the next chapter. It might be argued that effects take time so the value of y_t will rarely depend on the current x_t. However, if for example x_t and y_t represent yearly totals, the year covered by a single observation allows for short term influences of the x variable on the y variable.

6.3 The dependent variable—generalized regression*

So far we have concentrated on the linear model in the form

$$y_i = \beta_0 + \beta_1 x_{i1} + \beta_2 x_{i2} + \ldots + e_i,$$

where e_i is assumed to have zero mean, constant variance and often a normal distribution. This implies that the type of dependent variable we can consider is strictly limited. Thus if, in theory, the x's are continuous variables of infinite range then so must be the dependent variable. This rules out discrete variables. The zero mean of e prevents use of, for example, a positive exponential distribution as a model for dependent variables. For some situations we might also wish to consider a transformation of y on the left-hand side. To generalize the model to cover a wider range of dependent variables, it is best to seek to separate clearly the deterministic linear regression component from the random component. This can be done by reformulating the model using expectations. The above single equation can be re-expressed as three elements:

(a) The deterministic linear element

$$\eta_i = \beta_0 + \beta_1 x_{i1} + \ldots + \beta_q x_{iq}$$

(b) The distributional element

$$y_i \cap N(\mu_i, \sigma^2) \text{ with } E(y_i) = \mu_i$$

(c) The link element between these two

$$\mu_i = \eta_i$$

Extending the elements (b) and (c) to a wider range of distributions and links leads to a family of models called *Generalized Linear Models*. Let us consider some examples where this generalized formulation enables us to remove some of the limitations previously noted.

Example 1
The lifetime of a component is distributed as an exponential distribution.

(b) $$y_i \cap \text{Expn}\left(\frac{1}{\mu_i}\right) \text{ with } E(y_i) = \mu_i$$

The mean life of a specific component, y_i, depends linearly on a measure, x_i, of the stress applied to it. Thus we define

(a) $$\eta_i = \beta_0 + \beta_1 x_i$$

and the link that gives a shorter mean life for greater stress $x_i (x \geq 0)$ will be

(c) $$\mu_i = 1/\eta_i.$$

Example 2
The number of bubbles, r, per unit length in continuously produced glass is modelled by a Poisson distribution. Over a number of trials small changes are made to the operating temperature t. The mean number of bubbles decreases with increments in temperature. The model is thus

(a) $$\eta_i = \beta_0 + \beta_1 t$$
(b) $$r_i \cap P(\mu_i)$$
(c) $$\mu_i = e^{-\eta_i}$$

The form of the link here indicates, for positive β_1, a decrease in the mean number with temperature and also ensures a positive mean for the Poisson distribution.

Example 3
In an experiment on people's ability to detect a signal against a strong background noise, n trials were made at each of a number of different levels of background noise, x_i. The model was that the probability of detection was related to the noise. The model thus had:

The linear function
(a) $$\eta_i = \beta_0 + \beta_1 x_i.$$

The Binomial model for the r_i success in the n trials at noise level x_i

(b) $$r_i \cap \text{Bin}(n, \theta_i)$$

For the link a form is required that ensures that though η_i lies theoretically in $(-\infty, \infty)$ the value of θ_i lies in $(0, 1)$. This is achieved via the *logistic* transformation

(c) $$\eta_i = \ln \frac{\theta_i}{1-\theta_i} = \ln \left[\frac{Pr \text{ (success)}}{Pr \text{ (failure)}} \right]$$

or equivalently

$$\theta_i = \frac{e^{\eta_i}}{1+e^{\eta_i}}$$

Example 4
The quantities η_i in the previous example are called *Logits* and represent the natural logarithm of the *odds*, $\dfrac{Prob \text{ (success)}}{Prob \text{ (failure)}}$. A simple special case of this model is used in the modelling of qualitative data presentable in the form of Contingency Tables. For example suppose n_1 men and n_2 women are tested as to their ability to detect a particularly high musical note. The data would be tabulated as

numbers	detected	failed to detect	total
men	f_{11}	f_{12}	n_1
women	f_{21}	f_{22}	n_2

the probabilities being

probabilities	detect	fail to detect	total
men	p_1	$1-p_1$	1
women	p_2	$1-p_2$	1

In this case interest would focus on the difference, if any, between p_1 and p_2. Making use of the logistic transformation the model then focusses on whether the quantity

$$\Delta = \eta_1 - \eta_2 = \ln\left\{\frac{p_1}{1-p_1}\right\} - \ln\left\{\frac{p_2}{1-p_2}\right\} = \ln\left\{\frac{p_1(1-p_2)}{p_2(1-p_1)}\right\}$$

is zero or not. In terms of the regression model of example 3 this corresponds to

$$\eta_1 = \beta_0 + \Delta, \quad \eta_2 = \beta_0.$$

This type of model is referred to as the *Linear Logistic Model*.

6.4 Non-linear regression models

The development of the generalized linear model has extended the methodology of the linear model to many situations that in the past have been regarded as complex non-linear problems. Other models of non-linear form that fall outside the generalized linear model class can still be converted to linear form via a suitable transformation. We shall refer to both groups of models as *Linearizable Models*. Consider the following examples.

Example 1
A classical model in economics relates output (y) to capital (x) and manpower (z) by

$$y_i = \beta_0 x_i^{\beta_1} z_i^{\beta_2} e_i$$

This can be straightforwardly linearized by taking logarithms giving:

$$\ln y_i = \ln \beta_0 + \beta_1 \ln x_i + \beta_2 \ln z_i + \ln e_i$$

which is linear in the parameters $\ln \beta_0, \beta_1, \beta_2$. Note, however, that it is not linear in the original parameters $\beta_0, \beta_1, \beta_2$. Since $(\ln y_i, \ln x_i, \ln z_i)$ is simply a transform of the data this second form of the model is easier to study. The study of such models still needs care, since the original model is non-linear and the linearization does not avoid all the problems associated with non-linear models. The transformation used involves a much more radical step than the use of transformed regressor variables. This is clearly seen by examining the error term. If the transformed error term, $\ln e$, is to have zero expectation with variance σ^2, then the error variable e in the original model must have a more complex distributional form. For example if $\ln(e)$ is to have a normal distribution $N(0, \sigma^2)$, then e will have a lognormal distribution which has mean $e^{\frac{1}{2}\sigma^2}$. An implication of this is that, though $E(\ln y) = \ln \beta_0 + \beta_1 x + \beta_2 z$, as $E(\ln e_i) = 0$,

$$E(y) = \beta_0 x^{\beta_1} z^{\beta_2} e^{\frac{1}{2}\sigma^2}.$$

Hence the expectation of the output is not the deterministic part of the right-hand side. If the model after transformation has a simple structure the original model will involve statistical complications that must be noted and adjusted for.

Example 2
Consider a situation where an initial plot of the data on reciprocal graph paper shows a linear form. We could consider two possible models

A: $\qquad y_i = \mu_i + e_i, \ \mu_i = \dfrac{1}{\eta_i}, \ \eta_i = \alpha + \beta x_i, \ e \cap N(0, \sigma^2).$

This is a generalized linear model.

B: $\qquad \dfrac{1}{y_i} = \alpha + \beta x_i + e_i, \quad e \cap N(0, \sigma^2)$

This is a linear model for the transformed dependent variable, $1/y$. It will be seen that the difference in the two forms lies in whether the normal and constant variance distribution applies to y, as in A, or $1/y$, as in B.

The last example emphasizes the point that a correct formulation of non-linear models involves both the deterministic and stochastic components, together with the relationship between them.

Thus far our discussion of non-linear models has been concerned with models that can be linearized in various ways. This is not always possible. Consider the simple exponential growth model

$$y_i = \beta_0 + \beta_1 e^{\beta_2 x_i} + e_i$$

This cannot be transformed to linear form in any natural fashion. Table 6.2 gives a list of some non-linearizable models. It is sometimes proposed to linearize by using, for example, the deterministic model

$$\ln(y_i - \beta_0) = \ln \beta_1 + \beta_2 x_i$$

The occurrence of a parameter on the left-hand side, however, means that the model is still not linear in the proper sense of the word.

An important subset of non-linearizable models is of models that grow to a stable maximum, or minimum, level, technically a horizontal asymptote. These models are called *Asymptotic Growth Curves*. Table 6.2 lists these

Table 6.2 Asymptotic growth curves

For each curve y approaches α as x increases infinitely provided $\gamma < 0$

Simple modified exponential	$y = \alpha + \beta e^{\gamma x}$
Gompertz curve	$y = \alpha e^{(\beta e^{\gamma x})}$
Logistic curve	$y = \alpha/(1 + \beta e^{\gamma x})$

and indicates the relation of their shape to the parameters. Figure 6.1 illustrates a logistic growth curve.

The complexities arising with non-linear models can be seen from these few examples. There is the initial complexity that a model non-linear in the parameters is usually non-linear in the variables, so the relationship between dependent and regressor variables will take some complex form. As a consequence non-linearizable models are often difficult to distinguish from each other, particularly if data is only available for limited regions of the curve.

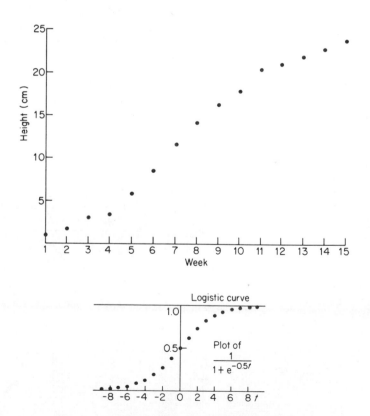

Figure 6.1 Data showing growth of a plant

The linear model has the elegant property that each variable and parameter has a separate influence on the dependent variable. Thus, if for $y = \beta_0 + \beta_1 x + \beta_2 z + e$ the variable x is increased by one, y is increased by β_1 irrespective of the values of other variables and other parameters. Even with a linearizable model, such as the output model considered at the beginning of this section, a percentage increase in capital, x, produces a percentage increase in output, y, that depends on β_1 but not on the other variables and parameters. However, if we consider the non-linear exponential growth model, the effect on y of changing x will depend on both β_1 and β_2. The effect of a change in the parameter β_2 will depend on the value of β_1.

Chapter 7

Progression models*

7.1 Introduction

The world is characterized by change. Many of the commonly used models therefore are those which are able to model change. We have already referred to regression models, such as

$$y_t = \beta_0 + \beta_1 x_t + \beta_2 t + e_t,$$

which provide an instantaneous, static, picture at time t. A natural development is to examine models which are inherently dynamic and which explicitly allow for the ways in which the past influences the present. For example the present value of the dependent variable y_t may depend on the current and immediate past values of the regressor variable, hence the model would be

$$y_t = \beta_0 + \beta_1 x_t + \beta_2 x_{t-1} + e_t.$$

Such models are called *Dynamic* or *Progression Models*. The purpose of this chapter is to study some of the more commonly used progression models. As a means of grouping these models we shall distinguish between models involving (a) discrete or continuous variables and (b) discrete or continuous time. The distinction by variable is one we have made before. Variables may be discrete counts $r = 3$ on the scale $r = 0, 1, 2, 3$ or continuous measurements $x = 7.346$ on the scale $(0, \infty)$. The idea of continuous or discrete time is a new consideration for us. In essence time is continuous and the way variables depend on time should be represented in a continuous fashion, e.g. $Y(t)$. $Y(t)$ is thus defined for all values of t. However, in practice three considerations frequently lead us to develop models that treat time as a discrete variable $t = 0, 1, 2, \ldots :$

(a) Though $Y(t)$ is continuous, observations are only obtained at regular times $t = 0, 1, 2, \ldots ,$ or sometimes at irregular times $t_0, t_1, t_2, \ldots .$ Only, say, $y(0), y(1), y(2), \ldots$ will be known and thus the model has to be designed to make use of such *Sampled Data*.

73

(b) It often occurs that though the fundamental variable, $Y(t)$, is continuous, what is observed is an aggregation or accumulation of all $Y(t)$ over fixed intervals of time. For example, though exports occur effectively continuously, these are recorded as monthly total exports, which form a discrete series.

(c) Sometimes the subscript t in y_t is used simply to denote the ordered sequence in which observations are obtained. For example r_t might be the number of faulty items found in the tth batch in a production run.

7.2 Models for continuous variables in discrete time

7.2.1 Moving average and autoregressive models

In this section we survey some of the major model forms in which the variables, Y_t, are observed at discrete time intervals. For simplicity we shall limit the description to situations involving equal, unit, time intervals. Consider a single variable Y observed at $t = 0, 1, 2, 3, \ldots$ In previous chapters we have often used models of the form

$$y_t = \ldots\ldots\ldots + e_t$$

where the e_t were independent, zero mean and constant variance. In these models each e_t, which, in this context, are often called *Innovations* or *Disturbances*, influences the corresponding y_t and no other. In, for example, the processes of economics, biology and continuous industrial manufacturing it is clear that the past influences the future so that the models used must allow each innovation, e_t, to have a continuing effect on the future. A very simple form of model that achieves this is the *Moving Average Model*. In these models the variable can be represented as a constant mean, or indeed any other known deterministic function of time, plus a weighted sum of the innovations. For simplicity consider observations in which the deterministic component has been subtracted. We write the general model as

$$y_t = e_t - \theta_1 e_{t-1} - \theta_2 e_{t-2} - \ldots - \theta_q e_{t-q}, \quad 0 \leqslant \theta < 1,$$

where the θ act as weights on past innovations. This model is called the *Moving Average of Order q* and denoted by MA(q). The structure of a MA(3) is illustrated in Figure 7.1(a). Using the standard assumptions about the innovations

$$E(Y_t) = 0,$$
$$V(Y_t) = \sigma^2 + \theta_1^2 \sigma^2 + \theta_2^2 \sigma^2 + \ldots + \theta_q^2 \sigma^2$$
$$= \sigma^2(1 + \theta_1^2 + \ldots + \theta_q^2) = \sigma_Y^2$$

A very important quantity in the analysis of progression models is the autocovariance of lag h defined by

$$\gamma_h = C(Y_t, Y_{t-h}) = E(Y_t Y_{t-h}) - E(Y_t)E(Y_{t-h}).$$

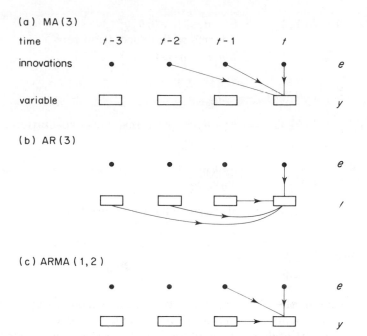

(a) MA(3)

time $t-3$ $t-2$ $t-1$ t

innovations e

variable y

(b) AR(3)

e

t

(c) ARMA(1,2)

e

y

Figure 7.1 Examples of the structure of AR, MA and ARMA models

This gives a measure of the relationship between Y_t and Y_{t-h}. When $h = 0$ this reduces to the variance at time t. Usually the variance is constant, as in the moving average model. It is common to 'standardize' the autocovariance to obtain the autocorrelation

$$\rho(Y_t, Y_{t-h}) = \frac{C(Y_t, Y_{t-h})}{\sqrt{V(Y_t)\,V(Y_{t-h})}}$$

For the MA(q) model we have

$$
\begin{aligned}
E(Y_t) &= E(Y_{t-i}) = 0, &\quad \text{all } i.\\
V(Y_t) &= V(Y_{t-i}) = \sigma_Y^2, &\quad \text{all } i.\\
E(e_t e_{t-i}) &= 0, &\quad \text{all } i.\\
E(e_{t-i}^2) &= \sigma^2, &\quad \text{all } i.
\end{aligned}
$$

These facts can be used to derive the autocovariance, γ, and autocorrelation, ρ, functions for moving average processes.

Example 1
Consider a MA(2) model

$$
\begin{aligned}
y_t &= e_t - \theta_1 e_{t-1} - \theta_2 e_{t-2}\\
y_{t-1} &= e_{t-1} - \theta_1 e_{t-2} - \theta_2 e_{t-3}\\
y_{t-2} &= \phantom{e_t - e_{t-1} -} e_{t-2} - \theta_1 e_{t-3} - \theta_2 e_{t-4}
\end{aligned}
$$

$$C(Y_t, Y_{t-1}) = E(Y_t Y_{t-1}) = E(-\theta_1 e_{t-1}^2 + \theta_1 \theta_2 e_{t-2}^2)$$
$$+ \text{terms with expectation zero}$$
$$= \gamma_1 \qquad\qquad = -\theta_1(1-\theta_2)\sigma^2$$

Hence $\rho(Y_t, Y_{t-1}) = \rho_1 = \dfrac{-\theta_1(1-\theta_2)}{1+\theta_1^2+\theta_2^2}$

$$C(Y_t, Y_{t-2}) = E(Y_t Y_{t-2}) = E(-\theta_2 e_{t-2}^2) + \text{terms with expectation zero}$$
$$= \gamma_2 \qquad\qquad = -\theta_2 \sigma^2$$

Hence $\rho(Y_t, Y_{t-2}) = \rho_2 = \dfrac{-\theta_2}{1+\theta_1^2+\theta_2^2}$

$$C(Y_t, Y_{t-3}) = E(Y_t Y_{t-3}) = \text{all terms with expectation zero}$$
$$= \gamma_3 \qquad\qquad = 0$$

and similarly for larger lags.

Hence $\gamma_k = \rho_k = 0$, $k > 2$.

Notice that in the above example the autocovariances and autocorrelations did not depend on t but only on the lag h. Processes in time whose statistical properties do not depend on actual time t are called *Stationary Random* (or *Stochastic*) *Processes*. Such processes will have constant mean and variance, and autocorrelations that depend only on the lag. It will also be noticed that for lags greater than two the MA(2) has autocorrelation zero. It is clear from the way this result occurs that it generalizes so that for an MA(q) process $\rho_h = 0$ for $h > q$.

A second important family of stationary random processes are the *Autoregressive Models*.

Example 2

In many economic situations the most important factor in determining a current value is the previous value. A natural model in such a situation is

$$y_t = \phi y_{t-1} + e_t, \quad \text{with} \quad 0 \leqslant \phi < 1.$$

Assuming stationarity of the random variable Y_t, $E(Y_{t-i}) = 0$, $V(Y_{t-i}) = \sigma_Y^2$ all t and i. Taking the variance of both sides, assuming that e_t is independent of Y_{t-1}, gives

$$\sigma_Y^2 = \phi^2 \sigma_Y^2 + \sigma^2$$

so $\qquad\qquad \sigma_Y^2 = \sigma^2/(1-\phi^2)$.

By repeated substititions

$$y_t = \phi(\phi y_{t-2} + e_{t-1}) + e_t$$
$$= \phi^2 y_{t-2} + \phi e_{t-1} + e_t$$
$$= \phi^2(\phi y_{t-3} + e_{t-2}) + \phi e_{t-1} + e_t$$
$$= \phi^3 y_{t-3} + \phi^2 e_{t-2} + \phi e_{t-1} + e_t.$$

Hence by continuing

$$y_t = \phi^h y_{t-h} + \phi^{h-1} e_{t-h-1} + \cdots + \phi e_{t-1} + e_t$$

Using this to obtain the autocovariance of lag h

$$E(Y_t Y_{t-h}) = E(Y_t Y_{t-h}) = E\left[(\phi^h Y_{t-h} + \phi^{h-1} e_{t-h-1} \cdots + e_t) Y_{t-h}\right]$$

so $\qquad \gamma_h = \phi^h E(Y_{t-h}^2) + \text{terms with expectation zero}$,

since Y_{t-h} is independent of innovations occurring after $t-h$. Hence

$$\gamma_h = \phi^h \sigma_Y^2,$$

and $\qquad\qquad \rho_h = \phi^h.$

Generalizing the above model we define the *Autoregressive model of order p*, AR(p) by

$$y_t = \phi_1 y_{t-1} + \phi_2 y_{t-2} + \cdots + \phi_p y_{t-p} + e_t.$$

The structure of an AR(3) is illustrated in Figure 7.1(b).

The term autoregression arises naturally since this is a standard regression model in which the current value is regressed on p previous values of the single variable y.

Multiplying the model equation through by y_{t-h} and taking expectations gives

$$\gamma_h = \phi_1 \gamma_{h-1} + \phi_2 \gamma_{h-2} + \cdots + \phi_p \gamma_{h-p} \qquad h > 0$$

so

$$\rho_h = \phi_1 \rho_{h-1} + \phi_2 \rho_{h-2} + \cdots + \phi_p \rho_{h-p}$$

Putting $h = 1, 2, \ldots, p$ in these equations and noting that $\rho_{-i} = \rho_i$ gives a set of simultaneous equations, the *Yule–Walker* equations, that relate $\phi_1, \ldots, \phi_p$ to $\rho_1, \ldots, \rho_p$, namely

$$\rho_1 = \phi_1 + \phi_2 \rho_1 + \qquad \cdots \cdots \qquad + \phi_p \rho_{p-1}$$
$$\rho_2 = \phi_1 \rho_1 + \phi_2 + \qquad \cdots \cdots \qquad + \phi_p \rho_{p-2}$$
$$\vdots$$
$$\rho_p = \phi_1 \rho_{p-1} + \phi_2 \rho_{p-2} + \qquad \cdots \cdots \qquad + \phi_p$$

Multiplying the model by y_t and taking expectations gives

$$\sigma_Y^2 = \phi_1 \gamma_1 + \phi_2 \gamma_2 + \cdots + \phi_p \gamma_p + E(y_t e_t)$$

Again using the model it is evident that

$$E(y_t e_t) = E(e_t^2) = \sigma^2. \text{ Thus, as } \gamma_i = \rho_i \sigma_Y^2,$$

we have

$$\sigma_Y^2 = \sigma^2 / (1 - \phi_1 \rho_1 - \phi_2 \rho_2 - \cdots - \phi_p \rho_p).$$

The autoregressive and moving average models have an interesting relationship. In the AR(1) example we showed that

$$y_t = e_t + \phi e_{t-1} + \phi^2 e_{t-2} + \cdots + \phi^{h-1} e_{t-h-1} + \phi^h x_{t-h}$$

It is evident that if h is large ϕ^h will become negligible and so the model is effectively an infinite moving average model with, in this case, $\theta_i = \phi^i$. A similar form of calculation shows that any $\mathrm{AR}(p)$ is equivalent to a $\mathrm{MA}(\infty)$. Conversely in the example on the $\mathrm{MA}(2)$ model the initial set of equations could be used to systematically eliminate e_{t-1}, e_{t-2}, etc. For example

$$y_t + \theta_1 y_{t-1} = e_t - (\theta_2 + \theta_1^2) e_{t-2} - \theta_1 \theta_2 e_{t-3}.$$

Repeating this process infinitely will relate y_t to all previous y_{t-i} and e_t. This model is an $\mathrm{AR}(\infty)$. Again this is a general result.

In exploring the choice of model for a prototype we often make use of the sequence of models $\mathrm{AR}(1)$, $\mathrm{AR}(2)$, ..., $\mathrm{AR}(k)$:

$$y_t = \phi_{1i} y_{t-1} + e_t$$
$$y_t = \phi_{21} y_{t-1} + \phi_{22} y_{t-2} + e_t$$
$$y_t = \phi_{k1} y_{t-1} + \phi_{k2} y_{t-2} + \ldots + \phi_{kk} y_{t-k} + e_t.$$

The coefficients ϕ_{11}, ϕ_{22}, ..., ϕ_{kk} are referred to as the *Partial Autocorrelation Coefficients*.

A natural extension of the moving average and autoregressive models is to combine them to create an *Autoregressive-Moving Average Model*, $\mathrm{ARMA}(p, q)$ defined by

$$y_t = \phi_1 y_{t-1} + \phi_2 y_{t-2} + \ldots + \phi_p y_{t-p} + e_t - \theta_1 e_{t-1} - \ldots - \theta_q e_{t-q}$$

The structure of an $\mathrm{ARMA}(1, 2)$ is illustrated in Figure 7.1 (c)

7.2.2 Transfer function models

The previous models involved a single variable. The natural extension is to involve several variables. Let us begin by considering situations which involve only two variables which have a natural causal relationship. In our previous terminology one is the dependent variable, the other the independent variable. In the common terminology of random processes one is the output variable, the other the input variable.

Example 3
A paper by Bondell, Gilmour and Sinclair studies models relating the input of rainfall, x_t, to the output of the overland flow of water, y_t, in a tropical forest. Two models proved to be very effective, the simplest

$$y_t = \beta_0 + \beta_1 x_{t-1} + e_t,$$

the other,

$$y_t = \beta_0 + \beta_1 x_{t-1} + \beta_2 x_{t-2} + e_t.$$

These models are like simple regression models in which the regressor variables are previous values, termed *Lagged Values*, of some causal quantity. In this example it is clear that variable x is causaly related to variable y in that x leads to y, $x \to y$.

The general lagged variable model, $L(r)$, can be written

$$y_t = \beta_0 + \beta_1 x_t + \beta_2 x_{t-1} + \, . \, + \beta_{r+1} x_{t-r} + e_t$$

Though models involving lagged values, which are known values at time t, look like simple regression models their behaviour is fundamentally different. Figure 7.2(a) and (b) illustrates.

(a) Regression involving time – static model $y_t = \beta_0 + \beta_1 x_t + \beta_2 t + e_t$

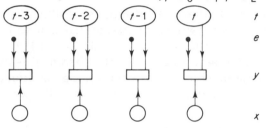

(b) Lagged variables model L(3)

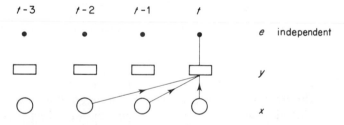

(c) Transfer function model L(3) + ARMA (1, 2)

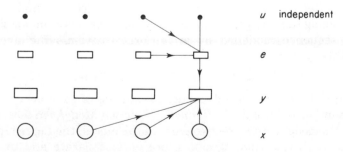

Figure 7.2 Examples of the structures of models

Example 4
Consider the deterministic part of the models

(a) $$y_t = 10 + 2x_t + e_t$$

(b) $$y_t = 10 + 2x_t + x_{t-1} + e.$$

Suppose x_t remains constant at $x_t = \mu$ until time $t = 10$ when it became thereafter $x_t = \mu + 1$, $t = 11, 12, \ldots$. The effect on model (a) would be an instantaneous increase of 2 at $t = 11$, Model (b) an $L(1)$; however, would increase by 2 at $t = 11$ with a further increase of 1 at $t = 12$. Thus to study (b) we must consider its dynamic behaviour through time.

Example 5
It is not unreasonable that in some situations the influence of a variable, x_t, an input variable, at time t will be greatest at that time and that its effect on future values, y_{t+k} of the output variable, y, will die away as time progresses and k increases. Conversely the current y_t is influenced by all past x, x_{t-k}, but the effect of each is a decreasing function of k. Thus if the decreasing effect is modelled by a multiplier λ^k, $0 < \lambda < 1$, a reasonable model of this situation is

$$y_t = \beta(x_t + \lambda x_{t-1} + \lambda^2 x_{t-2} + \ldots) + e_t,$$
$$y_t = \beta \sum_{k=0}^{\infty} \lambda^k x_{t-k} + e_t.$$

This is called the *geometric distributed lag* model.

The models in both the examples are special cases of the distributed lag model

$$y_t = \beta_0 + \beta_1 x_t + \beta_2 x_{t-1} + \ldots + e_t.$$

In the first example the number of parameters was restricted by only having the first few terms non-zero. In the second example, though there was an infinity of terms, only two parameters were needed since the parameters $\beta_0, \beta_1, \beta_2, \ldots$, were themselves modelled by

$$\beta_k = \beta \lambda^k.$$

A further extension of the general distributed lag model is possible by replacing the independent innovations, e_t, by a non-independent, stochastic sequence modelled by an autoregressive-moving average process, ARMA(p, q),

$$e_t = \phi_1 e_{t-1} + \ldots + \phi_p e_{t-p} + u_t - \theta_1 u_{t-1} - \ldots - \theta_q u_{t-q}$$

where the u_t are an independent sequence of random variables. Such a general model is called a *Transfer Function Model*. An essential feature of such a model is that the two parts of the model, the lagged input variables and the innovation structure, are quite separate and unrelated model components, as is shown diagrammatically in Figures 7.2(c) and 7.3.

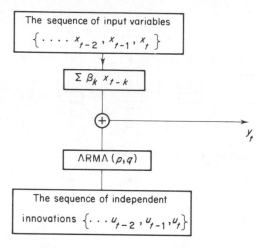

Figure 7.3 Transfer function models

7.2.3 Stochastic difference equation models

A natural extension of the Transfer Function model consists of situations where y_t depends not only on x_t and its past values but also on its own past values.

Example 6
Models of the growths of government expenditure, y, in developing countries have often shown the form

$$y_t = \beta_0 + \beta_1 x_t + \beta_2 y_{t-1} + e_t,$$

where x_t is a measure of the government's income. Such models can be conceptually derived. Assume a target expenditure based on income and add the constraint that change cannot necessarily take place as rapidly as would be required to meet the target. Thus y_t depends not only on income, x_t, but on the previous y_{t-1}.

A model of the form illustrated in the example is referred to as a *Stochastic Difference Equation Model*. The general form of such models is

$$y_t = \alpha_1 y_{t-1} + \alpha_2 y_{t-2} + \ldots + \alpha_p y_{t-p}$$
$$+ \beta_0 + \beta_1 x_t + \ldots + \beta_{q+1} x_{t-q} + e_t,$$

denoted by $D(p, q)$. As in the previous section, the e might not be independent but follow an ARMA process. It will be seen that this model is a mixture of autoregression with the transfer function model. The essential difference between this form of model and the transfer function form is the introduction of feedback, in the sense that changes in y_{t-1}, say, are directly

82

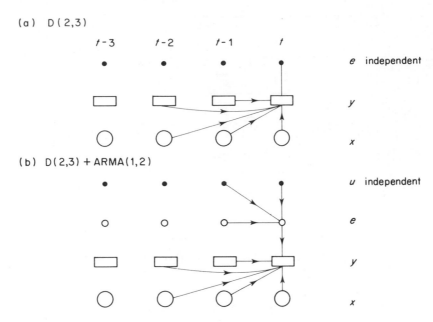

(a) D(2,3)

    $t-3$ $t-2$ $t-1$ t

e independent

y

x

(b) D(2,3) + ARMA(1,2)

u independent

e

y

x

Figure 7.4 The structures of stochastic difference models

fed back into the equation to influence y_t. Figure 7.4 illustrates the structural forms of these models.

7.2.4 Simultaneous equation models

The forms of progression model that we have thus far studied generalize to many variable situations. We consider prototypes where a set of variables, the *endogenous variables*, can be modelled in terms of another set of causal variables, the *exogenous variables*. The terms endogenous and exogenous, corresponding to output or dependent and input or regressor variables respectively in the previous simpler models, arose in economic literature and are now fairly widely used with the type of models discussed in this section. The generalization is shown symbolically in Figure 7.5. If, as in Figure 7.5(b) the endogenous variables, y, separately depend only on the exogenous variables, X, then the model is a *Multivariate Regression Model* which can be analysed in ways exactly analogous to those used for the ordinary multiple regression model of Figure 7.5(a). If, however, there is a causal link between the endogenous variables, the situation becomes much more complex. The reason for this can be seen from Figure 7.5(c). Not only do X_1, X_2 and X_3 influence Y_2 directly, they also influence it indirectly through Y_1. Similarly, Y_2 is not only influenced by the random e_2 but also by e_1, acting via Y_1. Thus Y_1 and Y_2 become correlated. Models of this general nature are called *Simultaneous Equation Models*.

(a) Multiple regression

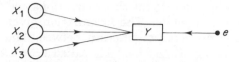

(b) Multivariate regression

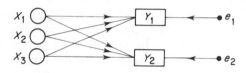

(c) Simultaneous equation model

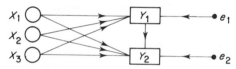

▢   Dependent, endogenous, variables

◯   Regressor, exogenous, variables

●   Random errors, disturbances, innovations

Figure 7.5 Regression of simultaneous equation models

Example 7

A simple classical model for a closed economy consists of three natural equations

$$\text{Consumption} = \alpha_0 + \alpha_1 (\text{Income}) + e_1$$
$$c_t = \alpha_0 + \alpha_1 y_t + e_{1t}.$$
$$\text{Investment} = \beta_0 + \beta_1 (\text{Income}) + \beta_2 (\text{lagged income}) + e_2$$
$$i_t = \beta_0 + \beta_1 y_t + \beta_2 y_{t-1} + e_{2t}.$$
$$\text{Income} = \text{Consumption} + \text{Investment} + \text{Government Expenditure}$$
$$y_t = c_t + i_t + g_t.$$

The first two equations express economic theory in simple linear form, the last equation states that all income is spent in one of the three ways indicated. The endogenous variables are c_t, y_t and i_t, the exogenous variables are g_t and y_{t-1}, since at time t y_{t-1} is assumed to be a known

84

causal factor. The above *Natural Form* of the model is represented in Figure 7.6(a), which shows the three equations and (b) which brings them together as a single model. The formulation given in the equations and diagram makes no clear distinction between the endogenous and the exogenous variables. Such a model is called a *Structural Model* or *Natural Model*. The structural equations arise naturally and directly from economic theory. However, if we are to use the model we shall need to emphasize how the exogenous variables effect the endogenous variables. So, as in a regression model, we should like to put the endogenous variables on the left-hand side of the equations, with the exogenous variables on the right-hand side. This is a laborious exercise in algebra. Rather than do this the main points become clear by examining the relationships as they appear on the diagrams of Figure 7.6. In part (c) of the figure the diagram has been sub-divided again to show how each separate endogenous variable is influenced by the exogenous variables only. Thus the first diagram shows that c_t is influenced by both g_t and y_{t-1} and also by both error variables, e_1 and e_2. The actual equation for this will take the form

$$c_t = \pi_0 + \pi_{11}g_t + \pi_{12}y_{t-1} + f_1,$$

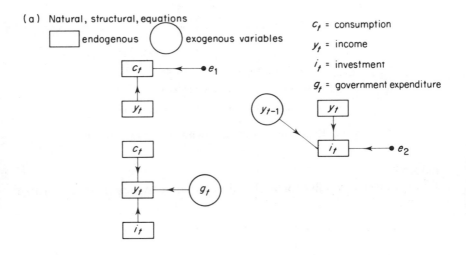

(a) Natural, structural, equations

□ endogenous   ○ exogenous variables

c_t = consumption
y_t = income
i_t = investment
g_t = government expenditure

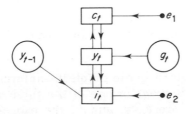

(b) Combined model

Figure 7.6 a and b

(c) Reduced form equations – showing original form

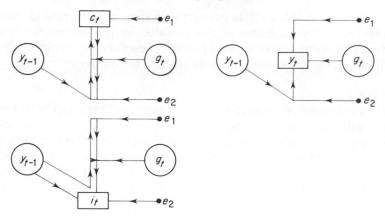

(d) Reduced form equations – simple form

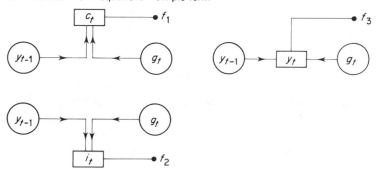

Figure 7.6 The simultaneous equation model of Example 7

where f is a composite of the error terms. This equation is obtained by using the second two structural equations to eliminate y_t and i_t from the equation of the first. Notice that the existence of two-way relations, e.g. $y_t \rightarrow i_t$, means that, for example, though y_{t-1} influences i_t directly, it will also influence it indirectly via its influence on y_t and c_t which in turn influence i_t. Similarly the errors will also have both direct and indirect effects on c_t and i_t. Models expressed to show clearly the total effects of exogenous variables on the endogenous variables are called *Reduced Form Models*. From Figure 7.6(d) it is clear that the three reduced form equations will take the general form

$$c_t = \pi_{10} + \pi_{11}g_t + \pi_{12}y_{t-1} + f_{1t},$$
$$i_t = \pi_{20} + \pi_{21}g_t + \pi_{22}y_{t-1} + f_{2t},$$
$$y_t = \pi_{30} + \pi_{31}g_t + \pi_{32}y_{t-1} + f_{3t}.$$

7.3 Models for continuous variables in continuous time

If a continuous variable $x(t)$ is varying with time, then its rate of change is represented in the language of the calculus by the derivative dx/dt. In building deterministic conceptual models of prototypes involving change, it will be obvious that many models involve derivatives.

Example 1
The simplest model of population growth assumes that $x(t)$, the (large) size of the population, can be treated as a continuous variable and that the rate of change, growth, of population dx/dt will depend on the rate of increase due to births and the rate of decrease due to deaths. Both of these will be proportional to the current size of the population $x(t)$. If λ and μ are the constant birth and death rates per person per unit time then the relation is

$$\frac{dx}{dt} = \lambda x - \mu x$$

$$\frac{dx}{dt} = kx \qquad (k = \lambda - \mu) \qquad \text{(A)}$$

Solving the equation by calculus gives

$$x(t) = x(0)\, e^{kt}, \qquad \text{(B)}$$

where putting $t = 0$ gives $x(0)$ as the initial size of the population. The population thus shows an exponential growth if $k > 0$ and an exponential decline if $k < 0$. Growth or decline obviously depending on whether the birth rate, λ, is greater or less than the death rate, μ.

The form of model, (A), in the example involving the derivative, we shall refer to as the *Natural or Structural Form*. Form (B) of the model is clearly equivalent to Form (A) but is more readily interpretable. We shall call such forms as (B) *Reduced* or *Solution Forms* since they arise by mathematical solution of the equations of the natural form. The conceptual building of models will lead invariably to the natural form. On the other hand empirical modelling, which is based on observing the way x varies in time will lead naturally to the solution form.

Table 7.1 gives a summary of the deterministic models commonly used to describe growth and decline. Each of these models involves the time path of one variable x.

Many physical models involve the interactions of several variables. An important set of these are those which involve *Through* and *Across Variables*. Through Variables are those that represent flows through systems, such as electric current, liquid current, heat flow, the velocity of objects. Across Variables are those that represent the influences that act across the system to produce the flow. Corresponding to the Through Variables listed we have, in order, voltage, pressure difference, tempera-

Table 7.1 Models of growth

Prototype	Natural model	Solution model	Name
1. Constant rate of change	$\dfrac{dx}{dt} = \delta$	$x = \alpha + \delta t$	*Linear growth*
2. Constant 'births'	$\dfrac{dx}{dt} = kx$	$x = \alpha e^{kt}$	Exponential growth
3. As 2 with steady emigration	$\dfrac{dx}{dt} = kx - \delta$	$x = \dfrac{\delta}{k} + \alpha e^{kt}$	Modified exponenital growth
4. As 2 with rate reduced by competition (dependent on x^2)	$\dfrac{dx}{dt} = kx - \eta x^2$	$x = \dfrac{k}{\eta}(1 + \alpha e^{kt})^{-1}$	Logistic growth

Note. In all the above solution forms α is a parameter arising from the solution.

ture difference and force. Fortunately the basic laws relating through variables to across variables are of the same form in electrical, hydraulic, thermal, and mechanical systems. In each type of system there are analogous elements corresponding to resistive, capacitative and inertial components (with one exception that thermal systems lack an inertial analogue). So for example if we use the electrical notation of i for the through variable (current) and v for the across variable (voltage) then resistive components (symbolized by —\/\/—) are well modelled by

$$v = Ri,$$

where R is a coefficient of resistance.

For inertial components (symbolized by —⟀—) such as an inductance or a mass

$$v = L\frac{di}{dt},$$

where L is a coefficient of inductance.

For a capacitative component (symbolized by —⊣⊢—) such as a capacitor or a spring

$$i = C\frac{dv}{dt},$$

where C is a coefficient of capacitance.

The fact that some components use di/dt and some dv/dt implies that when we model a set of linked components we will obtain models that involve higher derivatives such as d^2v/dt^2 (the rate of change of the rate of change of v). The tools of calculus enable the equations, called differential equations, of these models to be solved giving the corresponding solution models. It should be noted that the process of solution introduces additional parameters into the model, the number of additional parameters corresponding to the highest order of differentiation involved in the natural model. These parameters can often be found in terms of the initial state of the system.

The references by Doebelin and by Nicholson give a lead in to the current state of modelling with continuous variables and time.

Thus far we have referred only to deterministic models. Randomness may enter the picture in two fundamentally different ways:

(a) The randomness may appear due to measurement error in observing the situation. This will have the effect, for example, of adding an error term, $u(t)$, to the solution form, so in the introductory example the model would be

$$x(t) = x(0)\,e^{kt} + u(t), \qquad 0 \le t \le \infty.$$

If this process was observed at times $t = 0, 1, 2, \ldots$, then the observations could be written as

$$x_t = x_0\,e^{kt} + u_t$$

We thus have a standard non-linear regression model for discrete time and an analogous general model for continuous time. The continuous time error, $u(t)$, is referred to as a *White Noise Process* if $u(t_1)$ and $u(t_2)$ are independent for all t_1 and t_2. The analysis of deterministic solution form models with added white noise, is exactly analogous to the analysis of the discrete time situation. It should be noted that if $x(t)$ has any oscillatory behaviour there is a potential danger in observing $x(t)$ at regular intervals. This is illustrated in old films in which the wheels of fast moving trains appear to be rotating slowly backards. Here the time between frames, observations, is just less than is required for one revolution of the wheel. This phenomena is called *aliasing*.

(b) The modelling process becomes far more complex if the randomness is an inherent part of the underlying structure of the situation. In such cases random terms will appear in the natural form equations. Suppose in the population example there is an additive random rate of change due to a fluctuation in balance between immigration and emigration. The natural form will become

$$\frac{dx}{dt} = kx + u(t).$$

The analysis of such models is difficult and the mathematical tools for such analysis have only been developed in recent years.

7.4 Models for discrete variables in discrete time

For practical purposes there are two ways in which discrete variables arise in progression models. Firstly, they appear in problems involving discrete random counts $r = 0, 1, 2, \ldots$ such as models using Binomial or Poisson Variables. Secondly, they occur when the prototype includes a small number of possible states, see the reference by Howard.

Consider now a model appropriate to the situation where the variable is measured on the nominal scale and simply indicates a number of possible states. Usually the states are considered at a sequence of times $t = 0, 1, 2, \ldots$. However, to illustrate the possibility of using models, based on ordering in time, to apply in other situations involving natural ordering, we consider a geological example.

Example 1

If a core is taken from the earth in a region of sedimentary rocks a series of layers, or beds, is found. The beds may be classified as sandstone, shale, or limestone. As an aid to understanding the structure of such beds, it is useful to have a model that would enable one to predict the nature of the bed below one of a known classification. As this cannot be done with certainty, the prediction will have to be based on a probability model. Thus if we use subscripts 1, 2, and 3 to denote sandstone shale and limestone respectively then the probability of interest is

p_{ij} = Probability (bed of type i has bed of type j immediately below)

There are thus nine possibilities of transitions from a bed of class i to one of class j, $(i, j = 1, 2, 3)$. The quantities p_{ij} are termed transition probabilities. These nine probabilities can be presented as an array or matrix

| | | Lower bed | | |
		sand	shale	lime
	sand	p_{11}	p_{12}	p_{13}
Upper bed	shale	p_{21}	p_{22}	p_{23}
	lime	p_{31}	p_{32}	p_{33}

As all rocks in a sedimentary system are classified as one of the three types, it follows that the lower bed must be one of these. Hence the probabilities in each row must sum to one

$$p_{i1} + p_{i2} + p_{i3} = 1, \quad i = 1, 2, 3.$$

7.5 Models for discrete variables in continuous time

In this section we briefly illustrate some models that involve discrete variables and continuous time. At the simplest level the discrete random

90

variable indicates the state of some system that has a small number of possible states.

Example 1
A machine can be regarded as in one of two states at any time t, either a working state W, or a non-working state R, perhaps whilst being repaired after a failure. In general the probabilities of it being in either of these two states, $P_W(t)$ and $P_R(t)$, will vary with time depending on the time from initial start-up, $t = 0$, when the machine was working. The laws of probability will require that

$$P_W(t) + P_R(t) = 1.$$

The way $P_W(t)$ and $P_R(t)$ vary with time will depend on the processes which cause the machine to fail and those by which it is returned to working order. In a typical situation both $P_W(t)$ and $P_R(t)$ are composed of both constant terms and exponentially decaying terms so that the system stabilizes to constant probabilities, P_W and P_R, as t gets large.

Example 2
As a further example involving probabilities we shall briefly illustrate a *Queueing Model*. The mathematics of queueing theory is beyond the scope of this book. However, we can illustrate the nature of the model and make one fundamental point about probability based models by examining diagrams. Figure 7.7(a) shows a typical queueing situation with individuals arriving at random, queueing if the server is not free and spending varying amounts of time being served before departing. So for

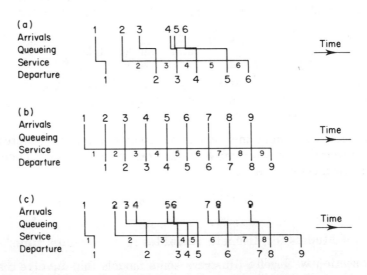

Figure 7.7 The nature of a queueing system

example, person 3 has to wait in the queue until person 2 has been served and departed. A comparison of diagrams (b) and (c) of the figure seeks to make the points that the randomness in such situations fundamentally influences their behaviour and that arguing on the basis of average times and deterministic behaviour is misleading. Thus in Figure 7.7(b) arrivals occur regularly at unit intervals and service takes this standard unit time. In this situation no queue ever forms. In Figure 7.7(c) the *average* times of arrival and of service are also one unit being 0.5 and 1.5 with equal probability. However, it is clear that queues often form. There can also occur gaps when no one is being served, these times cannot be made up in the averaging process so as time goes by it actually takes longer and longer to get through the system and queues contain more and more people. Thus though (b) suggests a highly efficient and balanced service system the effects of randomness revealed in (c) show that it is totally impracticable.

7.6 Models of complex systems

In the previous sections we limited the discussion to various specific models. If we are modelling complex situations for purposes of exploration we are led towards the development of highly complex models. These are usually obtained by application of what will be referred to as the *Principle of Modularity*. The complex prototype is reduced to a set of related *Modules*. Each module is modelled by some relatively simple model. These modules are then linked, with the possibility that the links themselves will require some modelling. If for example the analysis of the structure of employment in a city, one module might have as a variable the overall immigration of workers into the city. Another module might relate to the availability of housing. The interrelation between these two modules will require a modelling of how the availability of housing acts as one influence on attracting workers. This model in effect forms an additional linking module in the system. The modelling of such complex systems thus needs an initial modelling of the modules, *Submodels* or *Subsystems* as they are called. As the modules have to be linked together into complex structures, the variables used need to be those that enable a reasonable linkage. Also the models within the modules usually need to be relatively simple and compatible with each other. An initial study of the overall system is thus necessary. A description of the prototype by an *Influence Diagram* is a helpful means of beginning this. Figure 7.8 shows such a diagram.

Two forms of complex model in common use are the large models of national economies, used by most major nations, and the System Dynamics Models. These will be discussed very briefly:

(a) *Large economic models*
Developments in Economics and Econometrics in recent years have led to models of many parts of the economies of nations. Over the last twenty

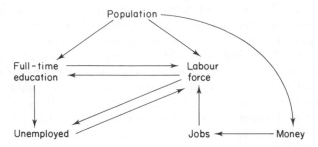

Figure 7.8 Influence diagram for a simple local unemployment model

years or so, many government agencies and large econometric and forecasting organizations have sought to use these as the modules of large-scale models for forecasting and policy analysis. Such models may involve several hundred variables and equations. The Treasury Macroeconomic Model of the United Kingdom Government involves over thirty modules, covering areas such as production, exports of goods and services, company profits and taxes, and debt interest. Naturally the modules and the flows between them are all basically financial. Most modules involve around twenty equations. The forms of the equations vary according to the variables being modelled and include almost all the forms of regression and progression models referred to in previous sections. The model has grown in complexity and sophistication over the years through a process of continual iteration. Many variables, that had to be treated as exogenous and externally determined in the early years, have become endogenous to the model as more aspects of economic behaviour have been modelled. In using these models there is a significant application of ad hoc adjustments for factors that are thought by 'experts' to be relevant, but which do not explicitly occur in the models. Such adjustments can often be carried out by adjusting a random term in the model, away from its zero statistical expected value, to some other economic expected value.

(b) *Systems dynamic models*
Most of the models under the heading of (a) have been built over long time periods from the bottom up, starting with detailed modelling of the modules. The approach that is usually referred to as *Systems Dynamics* starts at the top with a structure for the overall system based on the influence diagram. This technique, developed by Forrester, has been applied for example to study the behaviour of complex industrial, urban, global, and national systems. In such systems there are several types of variables linked together. For example, in an urban model the variables will involve people, jobs, buildings, finance, and space. There are thus subsystems, involving several modules, that relate to particular variables and their flows and change within the overall system. Individual modules,

however, will model how one variable is influenced by others. The design of the system seeks to make almost all variables endogenous and to include relevant goal seeking and decision aspects of the system. Most systems are constructed as a small number of major subsystems. For example, in a model of resource utilization (by W. W. Behrens in Meadows and Meadows) three subsystems are used—one for the use of resources, one for the technology of extraction and of substitutes, and one for the growth of substitutes. Each of these takes the form of a set of modules forming a loop of cause and effect relations. Thus demand is influenced negatively by cost and increased demand reduces the resources. Thus the three subsystems take the form of three *Causal Loops* which are linked by having cost as a common element. Such systems are sometimes referred to as *Multiloop Systems*. To obtain viable results with such models, without the massive resources that have gone into national economic models, the complexity of the multiloop structure is compensated for by using very simple relationships, as first approximations, within individual modules. For example if x_t is a particular variable at time t then the change over the small time interval to $t + k$ will be modelled by

$$x_{t+k} = x_t + r_t.k$$

The rate of change, r_t, may well depend on x_t itself, producing a form of feedback, or it may depend on some other variable, y_t, for example the usage rate of a natural resource will depend on the demand. Further relationships may involve the influence of delays that occur in many systems. For example the delays between investment in new research and the outcome of better resource extraction methods. A consequence of the linkage between loops, the delays and the feedback of variables on their own rates of change is that, though system dynamic models are composed of relatively simple modules, their overall behaviour is often highly complex and requires extensive computing for a thorough investigation by simulation. In general the models are too complex for a mathematical solution. It should be noted that the internal structure of these models is usually purely deterministic. Randomness, when it occurs, is usually generated in the simulation model by adding separate disturbance terms to particular variables. For example in a model of an inventory-production system the sales variable was simulated as a trend with added random disturbances.

Part III

Identification

Chapter 8

Prior information

8.1 Background

An essential component of all modelling is the study of what is already known relating to the problem. What information, what knowledge, do we posess prior to any gathering of data? What is our *Prior Information*? Clearly there is no shortage of information in this world, but what we need is relevant information. Initially it may not be clear what is and what is not relevant. What is kept and what is discarded will depend on our overall objectives in modelling and not just on the technical requirements of the methods we expect to use.

The form of prior information is specific to particular modelling projects. The aim of this chapter is to study various general forms of prior information.

8.2 Information about variables

The choice of which variables to use in a model is crucial. The neglect of an important variable can have some very strange effects when we come to apply the model. Some variables are essential to a model but many others may not be useful. It is essential to use the knowledge of the economist, the ecologist, the sociologist, etc., who most clearly understand the situation being modelled. We should first put the variables in groups of related quantities and order them, both within and between groups, in terms of likely importance. Other studies of these variables and their behaviour should be referenced. Relationships between variables and constraints on their values should also be carefully recorded. Constraints may take the form of absolute limits on variables, e.g. $X \geqslant 0$, $0 \leqslant Y \leqslant 100$ or possibly interrelationships, e.g. $X < Y$. Sometimes it may be useful to consider information from different but similar situations. For example, in a sales analysis a study of the distributions for sales data of a similar product may have revealed a lognormal distribution. This proves nothing about the product under study but it suggests directions for the investigation.

8.3 Information about data

A danger in modelling is that of devising and fitting sophisticated models with a high level of detail to sets of data that are too small or too unreliable. To have an observation of 2375 on its own tells us nothing about its precision—could it equally well have been 2377 or 2894? Thus an important component of the prior information needs to be information about the quality of the data available.

In many situations we must also consider the level of aggregation of the data. If for example we are modelling the growth of an insect population we have to ask how much detail we need to develop in the data. Can we gather data on the trees the larvae live on just considering them as trees, or do we need to have separate data on spruce, birch, etc? Similarly with time data do we need weekly or yearly totals? Such adding together or breaking down of data categories is called *Aggregation* or *Disaggregation*.

A further consideration is the availability of data. A model based on data from a special study may be of great interest but it may not be of future use, since the data necessary to adjust it to new situations may be too difficult or expensive to obtain.

8.4 Information about error and variation

The discussion so far has concentrated on the variables of the model. Thought must also be given to the sources of error and random variation that have to be modelled. We have emphasized in the discussion on models the importance of correctly specifying the relation between the deterministic and the stochastic part of the model: whether, for example, the error term in a regression model is additive or multiplicative. This is worthy of careful thought at this stage in the modelling process. We need to list the ways in which randomness might arise in the model. For example, if we are making measurements on an essentially deterministic situation, then for most measuring instruments we can initially assume that the error distribution does not depend on the magnitudes of the variables. Hence an additive constant variance error term is appropriate. If, however, we are examining a growth situation and the 'errors' are just the natural variation in the dependent variable then it is not unreasonable that the magnitude of the random variation will increase with the magnitude of the variable. This suggests an initial model involving a multiplicative error term.

Example 1
In a medical study the amount of body water, b, was derived by taking a proportion of the body weight, w, measurements being taken daily at times quoted to the nearest hour. There are three major potential sources of error.

(a) The error in measuring, w. Intuitively the use of hospital scales will give a relatively small error that should be independent of the magnitude of w, so an additive error, e, is a natural suggestion. The measured weight $\tilde{w}$ will thus equal $w + e$.

(b) The proportional factor, c, converting w to b, will vary in a random fashion from individual to individual. If the conversion factor c^* is used for all patients, then a suitable model would be

$$c = c^*(1+a),$$

where a is a zero mean random variable. Combining this with (a) suggests an estimate, $\hat{b}$, for b of

$$\hat{b} = c^* \cdot \tilde{w}.$$

The model relating b and $\hat{b}$ may be written as

$$\hat{b} = b + z.$$

Using the natural asumption of independence of a and e the error z, in estimating b, can directly be found as

$$z = c^*(e - aw)$$

and hence has mean and variance

$$(0, c^{*2}(\sigma_a^2 w^2 + \sigma_e^2)).$$

Notice that in this formulation the total error variance is not independent of w.

(c) The rounding of time to the nearest hour means that time is modelled by

$$t = T + \tau$$

where t is the actual time of measurement, T is the rounded time and τ is a random error which is taken as uniformly distributed in $(-\frac{1}{2}, \frac{1}{2})$.

Example 2

Perhaps the most commonly used and misused model is the simple linear regression. A basic assumption of the standard analysis of regression models is that the values of the regressor variables can be treated as known constants. In practice the values may often be measurements subject to error or they may be observations on some random variable. If the former, the standard methods only work if the measurement errors are relatively small compared with the random variation in the dependent variable. If the latter, then the problem is that the methods lead to reasonable results if the objective is to predict values of the dependent variable from given values of the regressor variables, but not in other applications. Thus the nature of error or natural variability in regressor variables is a topic of importance in the gathering of prior information. To see the importance of this point it is sufficient to realize that if a regressor variable, X, is a random variable then there is a whole new component to the model. We have to ask how the distribution of X is modelled and how the distribution of the dependent variable relates to this distribution.

A generalization from the last example is the fundamental question as to how data is obtained. Was it obtained by an experiment in which the environment of the prototype was controlled and variables carefully set and measured. Or was it obtained by simply observing an uncontrolled situation and measuring those things that it was practicable to measure. Even though we may be looking at the same variables in the two situations, there are clearly fundamental differences between (i) the models that will be appropriate, (ii) the ways these models may legitimately be analysed and validated, and (iii) the ways in which they will be treated and applied.

8.5 Information about models

In some areas of study there is a well-substantiated body of accepted theory which virtually provides the model to use. It may require sophisticated mathematics to get the desired form for the model, but the necessary basics are all provided. The journals of Applied Mathematics are full of models that are accepted as reasonable, but have never been tested on experimental data. The argument is that the premises have been thoroughly tested and therefore the deductions must hold. In other areas such as the Social Sciences there are many theories which are still very speculative and indeed there are often conflicting alternative theories that could lead to alternative models. Nonetheless their existence does suggest models that might be investigated.

A further consideration regarding the model is the level of detail required. For example, if the model is of a process that varies in time, do we wish it to describe variation from week to week or will it be adequate to describe year to year behaviour. The answer to this question may be forced by the data available but in general it will depend on the proposed application of the model.

It often occurs that the data available cover only a limited section of the range of possible values. For example in a regression, a regressor variable, x, may only be observed in some range (a, b). If the application of the model only requires a good fit in or very close to this range, then any model that gives a good fit to the data will be suitable. If, however, we wish to use the model to gain insight into the processes involved in generating the data, or to work in regions outside (a, b), then we need to seek a model that reflects more closely what is known about the prototype. For example over the range (a, b) the data may be adequately fitted by a linear or quadratic regression on x, with a constant term significantly different from zero. However, we may know from the nature of the prototype that the model should go through the origin, or that it should reach some stable maximum. In either case such information, together with the empirical evidence, suggests the need for a non-linear form.

8.6 Information about model parameters

The information we have about variables can sometimes be translated into information about the model structure or even its parameters. For example if we know that, irrespective of all other factors, an increase in X produces an increase in Y, then a requirement is imposed that the model behaves in this fashion. If we had a regression model;

$$y = \alpha + \beta x + \ldots + e,$$

this requirement becomes the requirement that $\beta > 0$ and implicitly that any sensible estimator $\hat{\beta}$ will satisfy $\hat{\beta} > 0$.

For models that will be in regular use by non-specialists, or by specialists in non-statistical or non-quantitative disciplines, it is important that the model is so formulated that its parameters have a clear practical interpretation. It may be that a more sophisticated model gives a better statistical description, but if the user has no intuitive feel for its structure or parameters then it is much more likely to be misused.

8.7 Probabilistic information about parameters*

So far we have stated our information in either descriptive fashion at one extreme or in mathematical fashion (e.g. $\hat{\beta} \geqslant 0$) at the other. It is sometimes possible to express our information in terms of probability statements that describe our uncertainties in a very explicit fashion.

Example 1
Consider the modelling of a production process that produces items in batches of ten. We would probably model the number of defectives by the Binomial model Bin $(10, p)$, where p is the probability of defectives occurring. Our prior information might tell us something about p. Consider three forms that the information might take

(a) Suppose we are dealing with new production machinery. We might decide beforehand that if in trials the machines could not guarantee $p < 0.01$ they would not be used, but subject to this any value of p was thought to be equally likely. We could then express this view by treating p as a random variable with p.d.f.

$$h(p) = 100, \qquad 0 \leqslant p \leqslant 0.01.$$

See Figure 8.1(a).
(b) If the machinery has been in use for some time we might use the daily production record to obtain the proportion defective each day. These could be plotted and a suitable model fitted. The form of model most

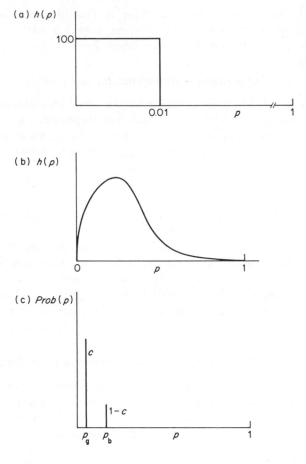

Figure 8.1 Prior distributions for a binomial probability

likely to give at least a first approximation to such data is the *Beta Distribution*

$$h(p) = \frac{1}{B(\alpha,\beta)} p^{\alpha-1}(1-p)^{\beta-1}, \qquad 0 \leqslant p \leqslant 1,$$

where $B(\alpha, \beta)$ is a quantity called a Beta function, that depends on α and β, but acts as a constant for the distribution of p. See Figure 8.1(b).

(c) It might be argued that, from a conceptual viewpoint, an analysis of the situation should be based on an extreme simplification in which p takes one of two values, one that represents the typical value of p, p_g, when the process is functioning properly; the other that represents the value of p, p_b, that would be typical when the process is not behaving properly. We could then let c be the probability of proper functioning.

Thus

$$Pr(P = p_g) = c, \qquad 0 \leqslant c \leqslant 1$$
$$Pr(P = P_b) = 1 - c.$$

See Figure 8.1(c). The parameter c could be found from an analysis of past production records.

In each of the three sections of the example we have treated a parameter, p, of one model as the random variable in the model of the prior information. The distribution of p is called the *Prior Probability* or *Prior Distribution*.

In situations where such prior probabilities are used the form and parameters of the prior may be based on (i) empirical evidence from previous experience, or (ii) on purely conceptual considerations that seek to express our view as to the actual shape of the prior distribution, or (iii) an eclectic approach combining (i) and (ii).

8.8 Probabilistic information about models*

In the same way that it can be helpful to formulate prior information about parameter values in probabilistic form, we can do the same for alternative possible models. Suppose for simplicity that there are only two models, M_0 and M_1, that we would consider. If we have no informed preference between them then we could say that there are 50:50 odds for the two models. If we have some previous experience giving some preference to M_0 we might give the 'prior odds' as 70:30, which we could write as *Prior Probabilities*, $Pr(M_0) = 0.7$ and $Pr(M_1) = 0.3$. Notice that our prior probabilities add up to one, so we ignore the possibility that some other model may be correct. This approach requires us to produce a list of possible models and to assign to them prior probabilities that express our beliefs and previous experience. It thus replaces the open ended question of 'what is the model?', by the question of choice 'which of these models?'.

8.9 Information about applications and criteria

The final area to be considered is the application of the model. Information about how the model is to be used can have a direct impact on the choice of model, on the way it is fitted and the way it is validated. Information about application thus leads directly to some of the considerations already discussed. Further information that should be considered concerns (a) the criteria by which a model will be judged, (b) how these might relate to the criteria by which it could be fitted, (c) whether the model will be part of a larger system and how it will relate to that system, (d) whether the fitted model will be used in a one-off situation or whether it will have steady use; if the latter then a more flexible model will probably be required to allow for some response to changes in the environment of the model.

Chapter 9
Conceptual identification

9.1 Introduction

The modeller begins with ideas, experience, and a desk full of relevant literature and data. The first task is to identify the model appropriate to the problem in hand. In this chapter we examine some conceptual approaches to identification. The modelling process will vary greatly from problem to problem and there are no general rules as to how to identify the appropriate model. Often the model arises naturally from the theory of the area of application. There are, however, a number of approaches and principles in model formulation that are of sufficient generality and usefulness for it to be worth devoting some space to considering and illustrating them.

9.2 Simplicity

Begin with the simplest forms of model that you can imagine. This is the natural starting point and it should be taken as the starting point. We shall discuss later, ways in which the data can be used to develop more realistic and sophisticated models from simple initial models. These methods often depend on the simplicity of that initial model. If the initial model is complex it may be very difficult to see where the changes should be made. Though most people would accept this principle in theory it is often abandoned for the sake of 'realism'. Take, for example, a financial planning model developed for the activities of a city's local government. An advisory committee produced initially a long list of all the factors required of a realistic model. The models, and the problems of getting all the relevant data, were of great complexity from the very beginning of the exercise. But yet some items in the list involved the expenditure of millions of pounds, others of only a few thousand. Though from the standpoint of realism these later items were thought necessary, attempts to include them as separate items in the basic model in fact greatly hindered the construction of a useful model. The complexity of reality does not imply the need for a complex model. The extent to which a model needs to be in any sense a reflection of the complexities of a prototype, depends on the use to which the model is to be put and the stage reached in the modelling exercise. Though the ultimate aim is to have a model that reflects 'reality', the requirement for initial simplicity is paramount.

It is useful to consider the concept of 'simplicity' a little further by listing some of the factors that effect the simplicity of a model:

(a) *The number of variables*

The simplicity of a model depends very largely on the number of variables that it takes into account. The principle of simplicity advocates keeping that number as small as possible. Where 'realism' insists that a large number of variables are non-negligible it may be possible to reduce this number of variables by aggregating them together in groups. This is done for example in economics when whole industries are grouped, as the sum of all the component firms, and in sales data, when annual data is used in place of the component weekly sales.

(b) *The number of parameters*

The number of variables does not alone determine the complexity of a model, since the same variable may appear more than once in the structure of the model. For example the variable x may appear just as x, but there may also be terms in x^2, e^{-x}, etc. within the model. The models might thus be

$$y = \theta x + e$$

or

$$y = \alpha + \beta x + \gamma x^2 + \delta e^{-\eta x} + e$$

The first model has only one explicit parameter, θ, the second has five, $\alpha, \beta, \gamma, \delta, \eta$. Thus the second factor influencing the complexity of a model is the number of parameters. The principle of simplicity advises us to use models with the minimum number of parameters. This particular case is called the *Principle of Parsimony*. The most parsimonious model is the one with the smallest number of parameters.

(c) *The model structure*

Simplicity in model structure relates to using standard distributions rather than less common ones, having independence rather than autocorrelation in the error terms, having linearity in variables and parameters rather than non-linearity.

Example 1

As an illustration of how a simple model assumption at the right place can lead to a quite sophisticated model structure, we use a model for the new product growth of consumer durables proposed by Bass. Let the variable T be a random individual's time to his first purchase of the durable. This will have p.d.f. $f(t)$ and c.d.f. $F(t)$. The probability of a first purchase at time t, $p(t)$, strictly $p(t)\,dt$ for a small interval dt, is the probability that $T = t$

conditional on no purchase being made before t, which has probability $[1 - F(t)]$. Hence the definition of conditional probability leads to

$$p(t) = f(t)/\{1 - F(t)\}.$$

The simple assumption underlying the model is that this probability will relate linearly to the proportion of the population of potential buyers who have already bought the product, which is $F(t)$. Thus

$$p(t) = \alpha + \beta F(t).$$

The parameter β relates to the willingness to imitate the buying behaviour of others, the parameter α represents the willingness to buy irrespective of other people buying.

Substituting from the definition of $p(t)$ gives

$$f(t) = \{1 - F(t)\}\{\alpha + \beta F(t)\}$$

which is a quadratic in $F(t)$. In data terms $f(t)$ is estimated by monthly sales figures during the period when repeat purchases can be neglected and $F(t)$ is estimated by the total sales to date. Bass and also Heeler and Hustad showed that this quadratic relation fitted well for many new durables in many countries.

Example 2
An example of using the assumption of linear relationships is provided by a model used in the diagnosis of diabetes, see the reference by Burghes and Borrie. The concentration of glucose in the blood and the concentration of hormones are mutually related. Both have a normal level which is disturbed in the test by an initial dose of glucose at time $t = 0$. Let g and h represent the deviations of glucose and hormone from their norms at time $t > 0$. The simplest model relates their rates of change linearly to the concentrations of both. Thus, using positive parameters α, β, γ, δ,

$$\frac{dg}{dt} = -\alpha g - \beta h$$

and

$$\frac{dh}{dt} = -\gamma h + \delta g$$

Differentiating $\dfrac{dg}{dt}$ with respect to time gives

$$\frac{d^2 g}{dt^2} = -\alpha \frac{dg}{dt} - \beta \frac{dh}{dt}$$

$$= -\alpha \frac{dg}{dt} + \beta \gamma h - \beta \delta g$$

$$= -\alpha \frac{dg}{dt} + \gamma \left(-\alpha g - \frac{dg}{dt} \right) - \beta \delta g$$

Hence

$$\frac{d^2g}{dt^2} + (\alpha + \gamma)\frac{dg}{dt} + (\alpha\gamma + \beta\delta)g = 0$$

This equation solves to show that g responds to the initial dose by showing a decaying oscillation. The period of this oscillation can be used to diagnose the occurrence of diabetes. We will return to this model later.

Example 3

It is sometimes possible to simplify the apparent structure of a model by transformations. As we have seen in section 6.4, the exponential growth model

$$y = \alpha z e^{\beta x},$$

where z is lognormally distributed, can be transformed to

$$\log y = \alpha' + \beta x + e.$$

This is a standard linear model with $\log y$ as the dependent variable. A problem occurs if z is not lognormal. If, as often happens only the deterministic part is specified,

$$y = \alpha e^{\beta x},$$

and the unspecified random component is not lognormal or is additive, then the transform might take

$$y = \alpha e^{\beta x} + z$$

to

$$\log y = \log(\alpha e^{\beta x} + z),$$

which is

$$\log y = \alpha' + \beta x + \log(1 + z e^{-\beta x}/\alpha).$$

Thus the transformation simplifies the deterministic part of the model at the expense of creating a complex random term which involves the parameters α and β. So unless the random component is of an appropriate form, transforming to get structural simplicity can be misleading, though such methods may be useful in empirical identification and in obtaining first approximations for parameter values.

9.3 Conservation

A principle that is used in many different areas of study is that of conservation. For example, the conservation of flows of Through Variables illustrated in Figure 9.1(a) where x is the flow into the junction and x_1 and x_2 are the output flows. The conservation law states that

$$x = x_1 + x_2.$$

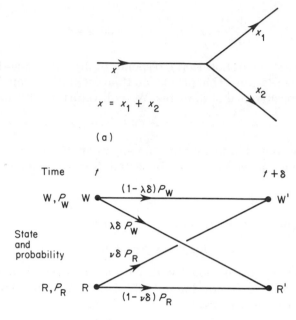

(a)

(b) Probability as a through variable

Figure 9.1 Conservation principles for through variables

The variable x could be:

  electric current along wires
  fluid flow through pipes
  mass flow of streams of charged particles
  flows of items along conveyor belts
  flows of people, e.g. in a single queue for two counters
  flows of money, e.g. income being spent or saved.

The equation $y_t = c_t + i_t + g_t$ of Example 7, section 7.2, illustrates the conservation of money.

In the context of this book it is worth while to examine the use of probability as a through variable obeying a conservation law. The conservation of probability arises from the fact that the probability associated with all possible events must be one. Thus in a plague each person in the original population can be classified, at time t, as infected $(I(t))$, or susceptible to infection $(S(t))$ or removed, by death or recovery and immunity, $(R(t))$. Hence for each person the probabilities of the three states must satisfy

$$Pr(I(t)) + Pr(S(t)) + Pr(R(t)) = 1 \text{ for all } t.$$

Example 1

A useful fact in developing models of probabilistic situations is that probability acts as a through variable between states of a system at different times. A machine at time t is in one of two states, either working, W, or under repair, R. A small time, δ, later it will again be either in states W or R. As the simplest model, assume that failures of a working machine occur at random with rate λ failures per unit time, or equivalently machines have an exponentially distributed working time of mean $1/\lambda$. Similarly assume repairs of a faulty machine take a time which is exponentially distributed with mean repair time $1/v$. These models imply that if δ is small

    Pr (a working machine fails in the interval δ) $\simeq \lambda\delta$
    Pr (a machine after repair commences working in the interval δ) $\simeq v\delta$.

As the machine can only be in W or R it follows that

    Pr (working machine is still working at $t + \delta$) $\simeq 1 - \lambda\delta$
and
    Pr (faulty machine is still under repair at $t + \delta$) $\simeq 1 - v\delta$.

Figure 9.1(b) represents this system as flows of probability through time between the two states W and R. The probability of a change from W at t to R at $t + \delta$ equals the probability of the transition W to R, $\lambda\delta$, multiplied by the probability of being in state W initially, P_W at time t. To simplify the illustration assume an equilibrium situation where the probabilities P_W and P_R, $P_W + P_R = 1$, of the two states are constant over time. In this case the flows shown in the figure must leave P_W and P_R unaltered.

Thus assuming conservation of probability

$$(1 - \lambda\delta)P_W + v\delta P_R = P_W,$$
and
$$\lambda\delta P_W + (1 - v\delta)P_R = P_R.$$

Hence $\lambda P_W = v P_R$, from either equation, which, as $P_W + P_R = 1$, gives

$$P_W = \frac{v}{v + \lambda}, \; P_R = \frac{\lambda}{v + \lambda}.$$

The non-equilibrium case, in which P_W and P_R vary through time, is dealt with in the same fashion but leads to more complex mathematical equations.

9.4 Accumulation

Accumulation Principles is a term that seems appropriate to the forms of conservation law and other rules that apply to Across Variables.

Example 1—*Additive accumulation*

Suppose that in Figure 9.2(a) the variables u, v, and w are across variables,

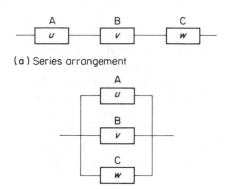

(a) Series arrangement

(b) Parallel arrangement

Figure 9.2 Accumulation

such as voltage drops across components, then the total, accumulated, voltage drop is

$$V = u + v + w$$

Example 2—Multiplicative accumulation
Suppose each block in Figure 9.2(a) represents a component that may either be operational or failed. The probabilities of the three components working, their *reliabilities*, are given by u, v, and w. The probability that the system as a whole works is

$$Pr \text{ (A and B and C all work)} = uvw$$

by the multiplication law of probability, assuming independence of operation. So for components in series a multiplicative form of accumulation operates.

In Figure 9.2(b) a system involving redundancy is represented by a parallel set of components. For the whole system to fail all the components must fail. The probability of this is

$$Pr \text{ (A and B and C all fail)} = \bar{u}\bar{v}\bar{w}$$

denoting $1 - u$ by $\bar{u}$. So again a multiplicative form of accumulation operates.

9.5 Parity

A further principle that applies to across variables, illustrated in Figure 9.3, is that the magnitude of the across variable, V, between two points, A and B, must be the same, on a par, for all paths between A and B.

Thus, from Figure 9.3

$$\text{Total } V \text{ by path 1} = \text{Total } V \text{ by path 2}$$
$$w = u + v$$

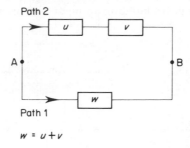

$$w = u + v$$

Figure 9.3 Parity principles for across variables

Table 9.1 illustrates the use of the principles of conservation, accumulation, and parity applied using the basic physical laws referred to in section 7.3.

9.6 Identity

A common source of terms in models is the statement that $X = X$—not very profound, but sometimes useful. For example, a model that involves two 'sectors' may, in its development, involve a consideration of $F_{AB.A}$ the flow of material from sector A to sector B viewed as an outflow from A. It might also consider $F_{AB.B}$ the inflow to B from A seen from B. Clearly, if no loss occurs, this is the same flow viewed from different directions so $F_{AB.A} = F_{AB.B}$. This type of term is sometimes called an equivalence relation but the term identity would seem a little more exact in its description; in fact equivalence would seem a more natural term to apply to what we have called parity to avoid confusion.

9.7 Modularity

A major problem facing the modeller is that produced by the complexity of real problems. As the number of variables increases the number of possible links and relationships increases dramatically. The most common way of dealing with this is to group elements of the model together into modules. Each module is then separately modelled. Interrelationships are then examined between modules and not between individual variables. In some areas of study rules, such as those derived from accumulation principles, enable a complex system of modules to be simplified. Figure 9.4 illustrates such a simplification process.

9.8 Alternatives

All modelling exericses should begin by an investigation of alternative ways of viewing the situation. It is very easy to slip too soon into just one

112

Table 9.1 Example of building an input–output system model

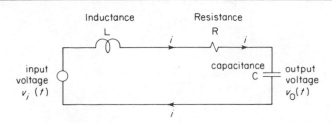

Conservation: The through variable, electric current, is conserved—i.e. is the same at all points in the circuit, as there are no junctions.

Accumulation: The sum of the across variables, voltage, for each component is the total across variable $= v_L + v_R + v_o$

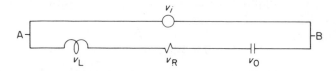

Parity: The across variable between A and B is the same by either path

so
$$v_i = v_L + v_R + v_o$$

From the 'laws' of section 7.3

$$v_R = Ri \qquad v_L = L\frac{di}{dt} \qquad i = C\frac{dv_o}{dt}$$

Hence
$$v_R = RC\frac{dv_o}{dt}, \ v_L = LC\frac{d^2v_o}{dt^2},$$

so
$$v_i(t) = LC\frac{d^2v_o}{dt^2} + RC\frac{dv_o}{dt} + v_o.$$

way of viewing a situation. This principle applies to many aspects of modelling, but for illustration we look at alternative ways of viewing a single variable in a problem and alternative ways of presenting a simple model.

Example 1

A record is kept of the occurrence of accidents on a stretch of road. Figure 9.5 shows a visual presentation of such a set of data. The original data consists of the series of times, T_1, T_2, . . . , at which the accidents occur. These times will originally be recorded as $T_3 = 14.50$ on Tuesday, 23 November 1982, etc. For simplicity of analysis these might be converted

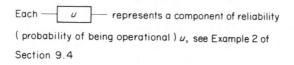

Each [u] represents a component of reliability

(probability of being operational) u, see Example 2 of

Section 9.4

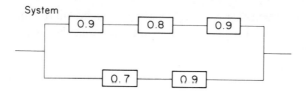

System

First simplification ($R_i = u_i\, v_i\, w_i$)

Second simplification ($1 - R = \bar{u}\, \bar{v}$)

Figure 9.4 Simplifying a modular system

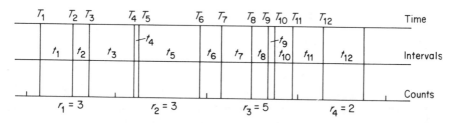

Figure 9.5 Series of events—alternative views

into times from some origin, e.g. $T_3 = 2574.89$ hours. By the nature of the situation the numerical values of T are always increasing. An alternative view is presented by examining the time intervals between events, i.e.

$$t_1 = T_2 - T_1,\ t_2 = T_3 - T_2,\ \ldots$$

We thus convert the data to get the new series, $t_1, t_2, \ldots$, and start considering how to model this sequence of t's. A study of models of this data may draw attention to a clustering of the events or long-term trends in accident frequency. This change of the data highlights some features of the situation at the cost of hiding other aspects. For example $t_{29} = 2.748$ hours tells us something of the gap between accidents 29 and 30 but nothing of whether these occurred in winter or sumer or day or night which was clear in the original data. Another way of looking at the same data is to divide the axis into uniform sections, say weekly periods and to count the number of events in each section. This gives a set of counts that would be modelled with a discrete distribution such as a Poisson.

Example 2
Consider a simple linear trend model

$$x_t = \alpha + \beta t + e_t.$$

Writing the mean at time t by

$$\mu_t = \alpha + \beta t$$

the model can be re-expressed by two equations, one describing how the underlying deterministic structure changes in time

$$\mu_{t+1} = \mu_t + \beta,$$

the other describes the observation process

$$x_t = \mu_t + e_t.$$

9.9 Analogy

A major source of inspiration in modelling over the years has been the use of analogy between one situation and another. The occurrence of analogous relations within the physical sciences was illustrated in section 7.3. With the extension of modelling outside the physical sciences such analogous models have been applied to a wide variety of situations. For example, in economics money has been treated as a through variable flowing through the economic system subject to conservation laws. The attractiveness of cities to people has been analysed by an analogy with gravitational attraction and the flows of traffic described on the basis of these 'Gravity Models'. The analogy that the attractiveness of city to shoppers will decrease with distance, as does gravitational attraction, is a reasonable one upon which to begin model building. To press the analogy to assume the 'inverse square' law that gravity satisfies would be to push the analogy too far.

As a further example it may be noticed that the form of the equation obtained in section 9.2 for the behaviour of glucose in the blood is exactly the same as that for the electrical circuit of Table 9.1. Here the analogy is in the behaviour but not in the structure that led to this behaviour.

9.10 Consistency

A further principle of conceptual modelling is that models should be internally logically self-consistent and also consistent with relevant prior knowledge that is not explicitly used in the model. The latter requirement can, however, be consciously ignored in seeking for simplicity. For example, we may know of long-term effects that we ignore in seeking a simple model for use in short-term decision making.

One aspect of internal consistency that is often helpful in the formulation of models, is that wherever possible the dimensions of terms should be consistent in equations.

Example 1

Suppose a mass, m, is moving with velocity v round a horizontal circle of radius r. That is to say it is whirled round on the end of a piece of string. It is reasonable that the tension in the string will depend on m, v, and r. Tension (F) is a force which Newton's laws give as a mass multiplied by an acceleration (say in metres per second per second). Thus the dimensions of the tension can be symbolized as

$$[\text{Mass}] \times [\text{Length}]/[\text{Time}]^2 = [M][L]/[T]^2.$$

The velocity of the mass (say in metres per second) has dimensions $[L]/[T]$. We require a relation

$$F = \text{some function of } m, v, \text{ and } r$$

The form that will give the dimensions on the right-hand side equal to those of the tension is:

$$[M] \times \left(\frac{[L]}{[T]}\right)^2 \times \frac{1}{[L]}.$$

This analysis thus identifies the model as being

$$F = \text{constant} \cdot mv^2/r.$$

Here the constant is just a dimensionless number, and in fact the dynamics of the situations gives the constant as unity.

This classical line of argument implies that, in situations based on physical laws, parameters should be dimensionless quantities. When one turns to much more indirect relationships, such as those often modelled in regression equation, then such dimensional consistency becomes impossible and the model parameters cannot be dimensionless. For example the monthly income (y) from sales of a spare part to a machine is modelled by a simple regression on the number of machines in use (n). Thus $y = \beta n + e$ and the parameter β is in units of money per month per machine.

Chapter 10

Empirical identification

10.1 Introduction

In conceptual identification we ignored the data. In this chapter we consider methods that are based at the other extreme of the modelling spectrum. We shall ignore our knowledge of the prototype and simply ask what the data alone can tell us. Experience suggests that this is a crucial but strictly limited exercise. It will provide a necessary understanding of the major features of the data. It is unlikely, however, to give a clear identification of the appropriate model. To do this we need to bring both prior knowledge and data together in a more eclectic approach to identification, which will be presented in the next chapter. In this chapter we build up some of the basic tools required for the data analysis part of the identification process.

The first group of tools are those that enable large sets of data to be condensed and summarized in ways that are clear and helpful. The numerical quantities that are used to provide summaries or indications of particular properties of data are called *summary statistics* or sometimes *indicators* according to the emphasis of their role. The most common summary statistics are those that indicate:

(a) the position of a set of data along the axis (section 10.2),
(b) the variability or spread of data (section 10.3),
(c) lack of symmetry in data (section 10.3),
(d) the strength of relationship between observations on different variables (section 10.4).

The second group of data analysis tools are those related to methods of data presentation. These are classified as:

(a) tabulation (section 10.5),
(b) semi-graphics (section 10.6),
(c) raw data graphics (section 10.7),
(d) indicator graphics (section 10.8).

116

10.2 Measures of position

The commonest measure of position, appropriate only for data of the form $x_1, x_2, \ldots, x_n$, is the *average* or *sample mean* $\bar{x}$ defined by

$$\bar{x} = \sum_{i=1}^{n} x_i/n.$$

This indicates the middle, the centre, the position of the data on the x-axis. For a set of data for which a frequency table has been constructed, so that f_i is the frequency of observations in the interval with x_i as its middle value, then the average of the original data is approximated by $\Sigma f_i x_i/n$. Table 10.1 illustrates the calculation.

If the data is ordinal or if we require, for interval-ratio data, a measure that is not too influenced by extreme observations, outliers, or lack of symmetry in the distribution, then the Median is used. The *Median* is defined as the middle item or value obtained when the observations are ordered. The value half-way between the two middle ones is used if there are an even number of observations.

For nominal data the *Mode* is the most frequently occurring observation. The Mode can also be applied as a measure of position to interval/ratio data by referring to the value or class with the highest frequency. Table 10.1 also illustrates these quantities.

10.3 Measures of spread and skewness

The basic quantity used for constructing measures of the spread or variability for interval/ratio data is the deviation of each observation from the average

$$d_i = x_i - \bar{x}.$$

If these quantities are large then the spread of the data is large. The average of the d_i^2 gives a suitable summary of the sizes of the d_i. Thus

$$s'^2 = \frac{1}{n}\Sigma d_i^2$$

is a measure of spread and is referred to as the *Sample Variance* (or the *Second Moment, m_2, about the mean*). s' is called the *Sample Standard Deviation*. A more usual definition is explained later (page 211).

For ordinal data the difference between the two extreme values indicates the spread of the observations.

In section 3.5(b) we introduced the idea of the percentile, x_p, which corresponded to a cumulative probability, p, expressed as a percentage. An empirical percentile can be analogously found for a set of data. The 25 % and 75 % percentiles, called the *Lower Quartile* and the *Upper Quartile* are particularly useful. Figure 10.5 shows their evaluation for a simple set of data. For more complex data the percentiles can be found from plotting the

Table 10.1 Examples of measures of position

(a) Data 7, 10, 11, 8, 10, 14, 9
  Sum = 69   Average $\bar{x} = 69/7 = 9.86$
  ordered data 7, 8, 9, 10, 10, 11, 14
  Median = 10

(b) Data on insect trapping (see section 14.2.1)

Individuals per trap r	Frequency (number of traps) f_r	rf_r	Cumulative frequency
0	6	0	6
1	8	8	14
2	12	24	26
3	4	12	30
4	3	12	33
	33	56	

Average $\bar{r} = 56/33 = 1.7$
Median = approximately 16th observation in order = 2
Mode = 2, observation with highest frequency

---

empirical cumulative distribution and reading off at the appropriate percentages. Figure 11.6 shows such a plot on special graph paper for which, for example, the lower quartile is about 9.4. The quartiles provide another simple measure of spread through the use of the difference between upper and lower quartiles, called the *Inter Quartile Range*.

Following the argument of section 2.5 an empirical measure of skewness is provided by

$$b^2 = \frac{m_3{}^2}{m_2{}^3},$$

where $m_3 = \frac{1}{n}\Sigma d_i{}^3$, is the *Third Moment* about the Mean.

10.4 Measures of relationship

Consider again the data of Figure 1.1. For data of such form it is often helpful to have some general measure of the linear relationship between the variables. A natural first step is to use not the observations but the corresponding deviations

$$d_{xi} = x_i - \bar{x}, \quad d_{yi} = y_i - \bar{y}$$

If there is a strong linear relation with a positive slope the paired (d_{xi}, d_{yi}) will tend to have the same sign. If the slope is negative they will tend to have opposite signs. If there is little linear relation between x and y then there will be little relation between the signs. This suggests that a measure of linear relationship is provided by

$$c_{xy} = \frac{1}{n}\Sigma d_{xi} d_{yi},$$

called the *Sample Covariance*, by analogy with the population quantity defined in section 2.5. As in the population case, it is useful to standardize this so that it does not depend on the variances. We thus define the *Sample Correlation* by

$$r = c_{xy}/s_x' s_y'.$$

For a perfect linear relationship, $y = x$, r is one, for a perfect negative relationship, $y = -x$, r is minus one. For essentially unrelated x and y, r will be close to zero and in general $-1 \leqslant r \leqslant 1$.

If a set of data $x_1, x_2, \ldots, x_n$ is in a natural sequence then we may be interested in the correlation between successive observations. We would organize the data in $(n-1)$ pairs $(x_1, x_2), (x_2, x_3), \ldots, (x_{n-1}, x_n)$. The correlation between these pairs is called the sample autocorrelation of lag 1, r_1. We might similarly define an autocorrelation of lag 2, by using the $(n-2)$ pairs $(x_1, x_3), (x_2, x_4), \ldots, (x_{n-2}, x_n)$, to obtain r_2. Similarly for lag 3, 4, etc.

10.5 Tabulation

An important point to remember in a preliminary identification exercise is that we are not interested in detail, we wish to know the shape of the forest not the details of the trees. To see the 'shape' of the data we use tabular or graphical presentation. To see 'shape' from a tabulation of the data we seek to know the magnitudes not the precise figures. Some rules suggested by Ehrenberg are well worth bearing in mind. These are:

(a) Round numbers to two significant or effective digits.
(b) Give averages for comparison.
(c) Put the most important comparisons in columns.
(d) Order rows and columns in a table by size (of average).
(e) Choose spacing and layout carefully.

10.6 Semi-graphics

Semi-graphics are presentations of data in which both a numerical and visual feeling for the data is given. The simplest of these is the stem and leaf display introduced, with other valuable approaches, by Tukey. The stem and leaf display shown in Table 10.2 illustrates how the numerical data can be presented to reveal the shape of the underlying distribution.

Table 10.2 A semi-graphic display of a set of data

(a) The Data. 18 observations 53.2, 114.3, 101.8, 94.0, 119.1, 108.0 125.4, 132.3, 135.9, 71.6, 88.5, 111.7, 114.6, 128.1, 128.8, 102.2, 96.9, 121.1.

(b) Ordered Rounded Data
53, 72, 88, 94, 97, 102, 102, 108,
112, 114, 115, 119, 121, 125, 128, 129, 132, 136.

(c) Stem and Leaf Display

tens	units
4	
5	3
6	
7	2
8	8
9	47
10	228
11	2459
12	1589
13	26

10.7 Raw data graphics

10.7.1 A single variable

Figure 10.1 shows a number of plots for observations on a single variable. The plots are largely self-explanatory. Plot (d) plots ordered deviations to the left of the median, $b = M - x$, against ordered deviations for data to the right, $a = x - M$. This provides a simple visual indication of the symmetry or otherwise of the distribution. Another plot showing this clearly is a plot of ordered data, $x_{(k)}$, against the rank position, k, as in (e).

10.7.2 Two variables

Given data, $x_1, \ldots, x_m, y_1, \ldots, y_n$, on two variables X and Y the plotting methods available depend on whether $m = n$ and, if it does, whether the observations have a natural pairing $(x_1, y_1) \ldots (x_n, y_n)$. If this latter case holds, then the data can be plotted as a scatter diagram as was done in Figure 1.1 for the data on straight line and road distances. If the data is unpaired a plot like Figure 10.2(a) may be helpful. Figure 10.2(b) shows some data from a situation showing a correlation between X and Y but no deterministic structure. The plot of figure 10.2(c) shows the same data with some added lines, that help to visualize the shape of the underlying distribution; the shapes drawn are called *Convex Hulls*. If the data is not paired but $m = n$ then the shapes of the distributions of X and Y can be compared using the ordered data paired by rank as in Figure 10.2(a).

An important application of the scatter diagram is in the choice of

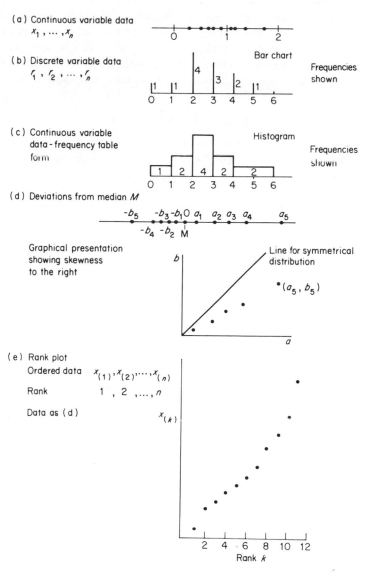

Figure10.1 Raw data graphics—a single variable

transformations to seek for linearity and symmetry. For example we have already seen that $y = \alpha x^\beta$ transforms to linearity with

$$\ln y = \ln \alpha + \beta \ln x$$

and $y = \alpha e^{\beta x}$ transforms to

$$\ln y = \ln \alpha + \beta x$$

The random part of these models has been neglected, since at the first part

122

of the identification stage we need simply an indication of the deterministic structure. The random component can later be investigated, using the residual analysis techniques of the next chapter. Thus the general form, either power curves or exponential growth curves can be identified by examining the linearity of scatter diagrams of the transformed data

$$(\ln x_i, \ln y_i) \text{ and } (x_i, \ln y_i).$$

Figure 10.3 illustrates such a plot using log-linear graph paper to avoid the numerical transformation. The plot suggests an exponential growth model to be appropriate, though approximate, for the given data.

As a further example Figure 10.4 presents data from a situation involving the growth of a new political party. The logistic model given in Table 6.2 provides the right shaped curve for this data (see Figure 6.1). From the model

$$y = \frac{\alpha}{1 + \beta e^{\gamma x}},$$

(a) Comparison of distributions – unpaired data

data $\quad x_1, x_2, \ldots, x_n; y_1, y_2, \ldots, y_n$

order $\quad x_{(1)}, x_{(2)}, \ldots, x_{(n)}$

$\quad\quad\quad y_{(1)}, y_{(2)}, \ldots, y_{(n)}$

Data shows similar shape of distribution on the left but y shows longer tail to the right

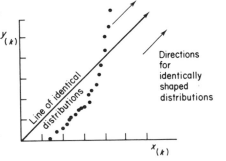

(b) A scatter diagram

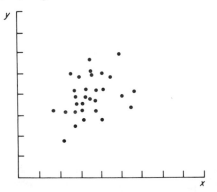

Figure 10.2 a and b

(c)

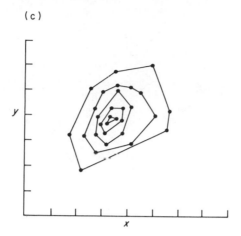

(d) Construction of a convex hull

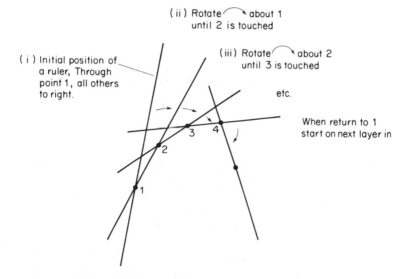

(ii) Rotate about 1
until 2 is touched

(i) Initial position of
a ruler, Through
point 1, all others
to right.

(iii) Rotate about 2
until 3 is touched

etc.

When return to 1
start on next layer in

Figure 10.2 Raw data graphics—two variables

and writing $p = y/\alpha$, it follows that

$$\ln\left(\frac{1-p}{p}\right) = \ln\beta + \gamma x.$$

The value of α can be tried at various values to find a linear plot.

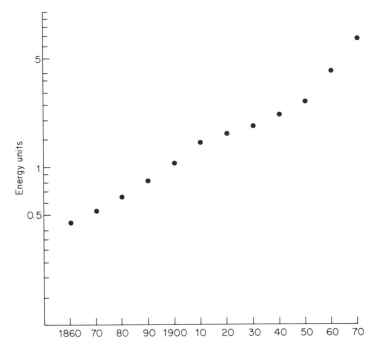

Figure 10.3 Log—linear plot of World Energy Consumption, 1860–1970

10.7.3 Many variables

When we turn to situations with more than two variables graphical analysis becomes more difficult. There is a practical problem in trying to visualize many dimensional data. With three-dimensional data, computer-produced stereoscopic displays are possible. For more than three-dimensional data two approaches are possible:

(a) Plot the data on scatter diagrams two variables at a time. Thus one can present a subset of the variables using all the observations on these variables. One immediate problem, if the number of dimensions is large is the sheer numbr of scatter diagrams that have to be drawn: for 20 variables there are $20 \times 19 = 380$ plots. There is a conseqent likelihood of getting apparent relationships in some plots by sheer chance or as a consequence of hidden relationships with non-plotted variables

(b) A second approach is to increase to three- or four-dimensional plots by using plotting symbols and colours based on putting some variables into classes as in a frequency table. With the increased availability of fast colour graphics, visual searches are becoming a useful tool of empirical modelling.

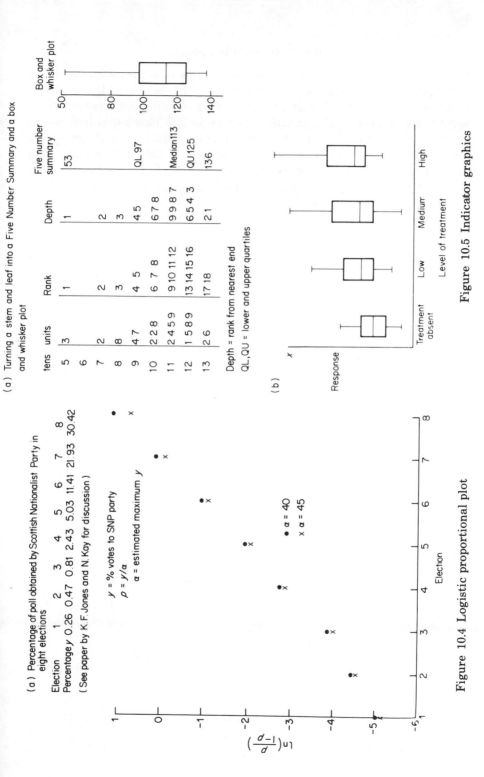

(a) Percentage of poll obtained by Scottish Nationalist Party in eight elections

Election	1	2	3	4	5	6	7	8
Percentage y	0.26	0.47	0.81	2.43	5.03	11.41	21.93	30.42

(See paper by K.F. Jones and N. Kay for discussion)

y = % votes to SNP party
p = y/α
α = estimated maximum y

● α = 40
x α = 45

y-axis: $\ln\left(\frac{p-1}{p}\right)$ x-axis: Election

Figure 10.4 Logistic proportional plot

(a) Turning a stem and leaf into a Five Number Summary and a box and whisker plot

tens	units	Rank	Depth	Five number summary
5	3	1	1	53
6				
7	2	2	2	
8	8	3	3	
9	4 7	4 5	4 5	QL 97
10	2 2 8	6 7 8	6 7 8	Median 113
11	2 4 5 9	9 10 11 12	9 9 8 7	QU 125
12	1 5 8 9	13 14 15 16	6 5 4 3	136
13	2 6	17 18	2 1	

Depth = rank from nearest end
QL, QU = lower and upper quartiles

(b)

Response vs Level of treatment

Treatment absent   Low   Medium   High

Figure 10.5 Indicator graphics

10.8 Indicator graphics

Our final series of illustrations are based on seeking for indicators that, when plotted, will show particular aspects of the data in a clear fashion. Box-and-whisker plots, Figure 10.5(a), give a useful indication of position, spread, and skewness. For example, Figure 10.5(b) shows that increasing the level of treatment has an influence not only on the level of response but also on its spread and skewness. It is seen that the plot indicates the median, the quartiles, and the extremes of the data. For large data sets the 5 and 95 percentiles are sometimes used in place of the extremes.

The idea of indicators will be explored further in section 17.6.

Chapter 11

Eclectic identification—graphical techniques

11.1 Introduction

In the two previous chapters we considered how to construct and identify models, either by using prior information and knowledge or by allowing the data available to speak for itself. In some situations the use of only one of these approaches may be justified, but in general both approaches should be used since each is clearly inadequate alone. So far we have considered their separate use with the various techniques and approaches appropriate. The objective of this chapter is to consider some more eclectic approaches in which we try to combine the conceptual and empirical ways of thinking. This combination is usually based on the common experience that we have prior knowledge which suggests some particular model which should be tried first. For distributional models, the normal distribution or a model related to Bernoulli or Poisson processes might be suggested by consideration of the prototype. In using regression, past experience of similar situations may suggest which variables to consider first. In a progression situation, the common occurrence and simplicity of the autoregressive model might suggest it as a starting point. This chapter will concentrate on ways of examining the data in the light of such initial models. It will sometimes be necessary to assume that this initial model has already been fitted to the data. The methods for doing this are discussed in the next part of this book. As we have mentioned before, the process of modelling does not in practice follow the sequential rules of chapters in a book. We often need to iterate the process several times. We shall thus assume that where necessary we have estimates of the numerical values of parameters of the initial model from the data under study.

11.2 Probability plotting

This section is concerned with the question of how well a given distribution describes a set of data, for example, how well an exponential distribution describes the distances between faults on a wire, or how well a normal distribution models the errors made in forecasting. The simplest methods of examining the closeness of distributional models to data are visual and are described as *Probability Plots*. Some of these plots can be made on ordinary graph paper, others require special graph papers. A list of such special papers is given at the end of this section. The aim of this section is to illustrate a number of different probability plotting methods.

11.2.1 Plots based on cumulative distribution functions, $F(x)$.

Example 1—Using ordinary graph paper
We have seen that the probability density function of the intervals between points in a Poisson process is exponential:

$$f(x) = \lambda e^{-\lambda x}$$

The cumulative distribution function, c.d.f., for this distribution is

$$F(x) = Prob(X \leqslant x) = 1 - e^{-\lambda x},$$
$$\text{Hence } Pr(X > x) = 1 - F(x) = G(x) = e^{-\lambda x}.$$

Table 11.1 gives a frequency table of times between arrivals, for which an exponential distribution might be appropriate. Estimates $\hat{F}(x)$ and $\hat{G}(x)$ of $F(x)$ and $G(x)$ can be obtained from the table by replacing the probabilities $Pr(X \leqslant x)$ and $Pr(X > x)$ by the corresponding accumulated relative frequencies. This is done to give $\hat{G}(x)$ in the table. The quantity $\hat{F}(x)$ is the *Cumulative Relative Frequency* sometimes called the *Empirical c.d.f.*

A plot of $\hat{F}(x)$ against $F(x)$ will thus give approximately a straight line if the model is correct. A slightly simpler plot can be obtained taking logs and plotting $-\ln \hat{G}(x)$ against x, since $-\ln(1 - F(x)) = \lambda x$. This plot is illustrated in Figure 11.1. Notice that we do not need an estimate of λ to

Table 11.1 Times between arrivals

Interval	Frequency	Reverse cumulative frequency	$1 - \tilde{F}(x)$ $= \tilde{G}(x)$	x	$-\ln \tilde{G}(x)$
0—5⁻	86	200	1.00	0	0
5—10⁻	41	114	0.57	5	0.56
10—15⁻	34	73	0.365	10	1.01
15—20⁻	16	39	0.195	15	1.63
20—25⁻	8	23	0.115	20	2.16
25—30⁻	8	15	0.075	25	2.59
30—35⁻	1	7	0.035	30	3.35
35—40⁻	3	6	0.030	35	3.51
40—45⁻	2	3	0.015	40	4.20
45—	1	1	0.005	45	5.30

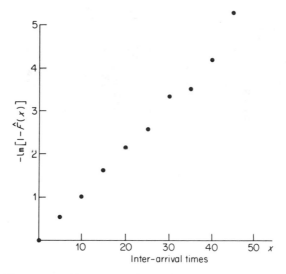

Figure 11.1 Exponential plot for data of Table 11.1

obtain this plot and in fact the slope of the line gives an estimate of λ. Examining Figure 11.1 indicates that the exponential model gives a reasonable fit to the data.

Example 2—Using special graph papers—Continuous distributions
In many situations a natural question is whether a Normal distribution is appropriate. To illustrate the approach consider the data in Figure 11.2(a) which gives frequency, the cumulative frequency and the empirical c.d.f., $\hat{F}(x)$. The c.d.f. for the Normal distribution does not have a simple functional form. However, it is possible to devise a transformation to produce a straight line plot of $F(x)$ against x. To avoid the mathematical complexities of this exercise special graph paper is available that transforms the axes to produce a straight line form. This paper is called *Normal Probability Paper* or, often, just *Probability Paper*. Figure 11.2(b) shows the data of (a) plotted on probability paper.

Figure 11.3 shows two probability plots in which the points deviate from a straight line. Let us consider briefly the interpretation of these deviations. In Figure 11.3(a) the points at the right end of the plot have larger values of x than a straight line plot would give thus the distribution of x would appear to have a longer tail to the right than the model used as the basis for the probability paper. Conversely in Figure 11.3(b) the points on the right are closer in than the linear position, indicating a very short tail to the right of the distribution.

There are a range of plotting papers for commonly used distributions, e.g. χ^2, Gamma, Weibull, Extreme value. These all follow the method illustrated for the normal distribution. The Poisson and Binomial distri-

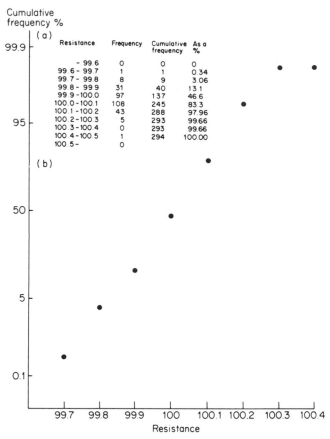

Figure 11.2 A normal probability plot

butions, however, have plotting papers of rather a different form. The next examples illustrates the testing of a Poisson model using the appropriate plotting paper.

Example 3—Using special graph papers—Poisson paper

A series of traps were set out in line across sand dunes and the numbers of different types of insects caught in a fixed time were counted. The numbers of traps containing the specified number of insects for two different taxa were as in Table 11.2(a). This data would be modelled by a Poisson distribution if the insects fell into the various traps at random. If, however, some of them tended to concentrate in certain regions of the dunes then some traps would pick up more than suggested by the Poisson model and some less. The data can be graphically studied using *Poisson Probability Paper*, Figure 11.4. The vertical axis gives the probabilities, from the Poisson distribution, of observing at least c invertebrates in a trap. This is estimated by the corresponding proportion which is shown as a percentage in Table 11.2(b). The value of c is given in the plot by the curved lines and

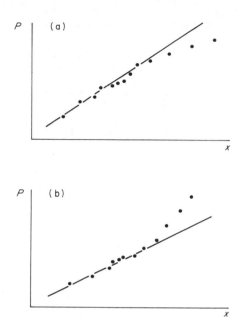

Figure 11.3 Deviations from linearity

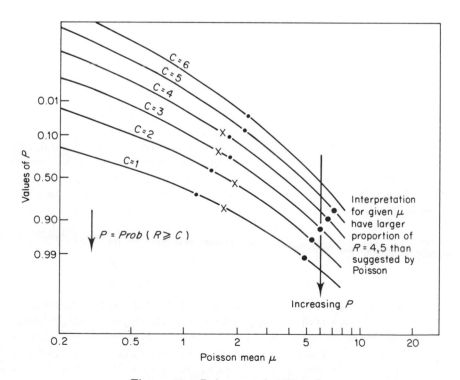

Figure 11.4 Poisson probability plot

Table 11.2 Invertebrates trapped

(a) Frequency of traps	Individuals per trap (33 traps)							
Taxa	0	1	2	3	4	5	6	Totals
Staphylinoidea (Rove beetles)	10	9	5	5	1	2	1	33
Hemiptera (Leaf hoppers)	6	8	12	4	3			33

(b)		Percentage of traps with at least c individuals						
Taxa	$c =$	0	1	2	3	4	5	6
Staphylinoidea		100	69.7	42.4	27.3	12.1	9.1	3.0
Hemiptera		100	81.8	57.6	21.2	9.1		

the horizontal axis gives the mean μ for the distribution. Thus perfect data would show as a vertical line. A concentration would show up as a deviation from the vertical, sloping up to the right. It is seen that the Hemiptera data is reasonably vertical but the Staphylinoidea shows a tendency to cluster. So a Poisson model is suggested for the Hemiptera.

11.2.2 Plots based on order statistics*

Sometimes if we try to use the c.d.f. approach of the previous section we find that there is insufficient data, so we seek methods that make use of every observation. To do this consider the ordered observations denoted by $x_{(1)}$, $x_{(2)}, \ldots , x_{(n)}$, the smallest observation being $x_{(1)}$, the largest $x_{(n)}$. The subscripts $(1), \ldots ,(n)$ denoting the rank order. Even if the original observations were normal the ordered observations individually will not be. For example $x_{(1)}$ is an observation on the random variable $X_{(1)}$ which will have a mean, $\mu_{(1)}$, smaller than μ and its distribution will have a long tail to the left and a short one to the right, for there is very low probability of the smallest observation being greater than μ. The expected values of the random variables $(X_{(1)}, X_{(2)}, \ldots , X_{(n)}$, denoted by $\mu_{(1)}, \mu_{(2)}, \ldots ,\mu_{(n)}$, are called *Rankits*. If X is $N(0, 1)$ they are referred to as *Normal Scores* (tabulated in Neave). If we now plot the pairs

$$(\mu_{(1)}, x_{(1)}), (\mu_{(2)}, x_{(2)}), \ldots , (\mu_{(n)}, x_{(n)}),$$

we shall be plotting each $x_{(i)}$ against its expected value so the points will be scattered about a straight line provided the underlying distributional assumption is correct. Figure 11.5 illustrates a Normal Scores plot. If tables of Normal Scores, are not readily available an equivalent plot can be obtained using Normal Probability Paper. If $F(x)$ is the normal c.d.f. it

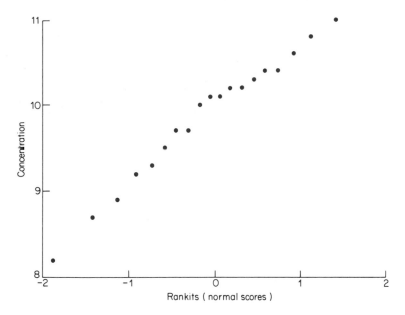

Figure 11.5 Normal plot, concentrations (g/litre) (given in Chapter 13.3.1) using normal scores

can be shown that, approximately, $F(\mu_{(i)}) = \dfrac{i}{n+1}$, thus a plot of $x_{(i)}$ against $\dfrac{i}{n+1}$ will give approximately a straight line on Normal Probability Paper. Figure 11.6 gives a plot of the data on such paper.

11.2.3 Plots based on simulation*

The previous methods require convenient mathematical forms or, special graph papers or tables. If such are not available it is often possible to make use of computer simulation. This possibility relies on two factors. Firstly most computers have simple instructions that enable them to generate *Pseudo Random Numbers*, i.e. sequences of numbers that appear to show properties of randomness. These are numbers from a uniform distribution in (0,1). Secondly, in theory all distributions can be obtained from transforming a uniform distribution, see Table 11.3(f). The transformation may be difficult to handle but often there exist quite simple ways of using random numbers to obtain other distributions. Table 11.3(a)–(e) give some examples. The speed at which computers operate enables these two facts to be combined to simulate large numbers of pseudo observations from any assumed model. These can then be used as the basis of comparison with the

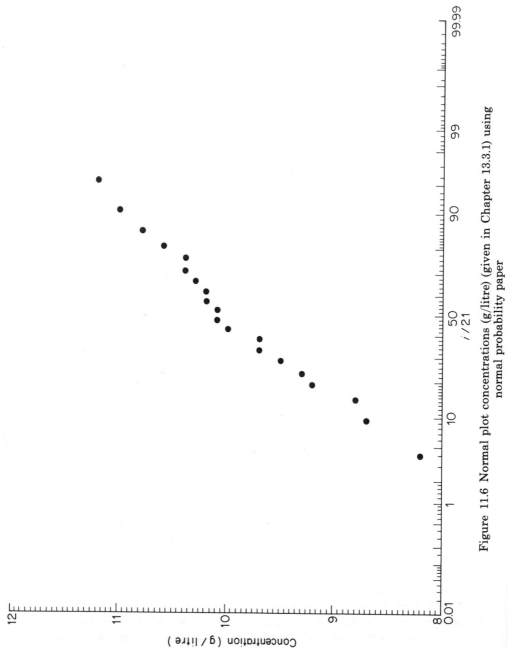

Figure 11.6 Normal plot concentrations (g/litre) (given in Chapter 13.3.1) using normal probability paper

Table 11.3 Simulation of standard distributions

(a) *Uniform Distribution*
$$f(u) = 1 \quad 0 \leqslant u \leqslant 1 \quad \text{zero otherwise}$$

Given by Random Number Generator on most computers, e.g. $u = \text{RND}$.

(b) *Binomial Distribution*
$$Pr(r) = \binom{n}{r}\theta^r(1-\theta)^{n-r}, r = 0, 1, \ldots, n; 0 \leqslant \theta \leqslant 1$$

Construct n indicator variables
$$z_i = \begin{array}{l} 1 \text{ if } u_i \leqslant \theta \\ 0 \text{ if } u_i > \theta \end{array}, \quad u_i \text{ from (a).}$$

$$r = \sum_{i=1}^{n} z_i.$$

(c) *Exponential Distribution*
$$f(x) = \lambda e^{-\lambda x}, \quad x \geqslant 0, \lambda > 0.$$

$$x = -\frac{1}{\lambda} \ln u.$$

(d) *Normal Distribution*
$$f(x) = \frac{1}{\sqrt{2\pi}} \exp -\tfrac{1}{2}x^2$$

$$\text{(i)}x = \sum_{1}^{12} u_i - 6$$

$$\text{(ii) } y \cap N(\mu, \sigma^2) \quad y = \mu + \sigma x$$

(e) *Poisson Distribution*
$$Pr(r) = e^{-\mu} \frac{\mu^r}{r}!, r = 0, 1, 2, \ldots,; \quad \mu > 0$$

Generate exponentials with $\lambda = 1$ as in (c). Successively add until $\Sigma x < \mu$. Number of exponentials used is then $r + 1$.

(f) *General Distribution, $f(x)$*
Construct cumulative distribution function $F(x)$. Generate u_i as in (a) and find corresponding x_i. The x_i will be observation on the $f(x)$ distribution.

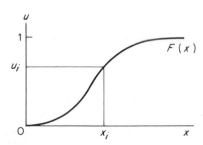

$$f(x) = \alpha x^{-(\alpha - 1)}, \quad x \geqslant 1, \alpha > 1$$
$$x = (1/u)^{1/\alpha}$$

(g) *Pareto Distribution*

observations. To see the theory of the method let the ordered data be denoted by

$$x_{(1)} \; x_{(2)} \; \cdots \; x_{(n)}$$

We now simulate pseudo observations, z, from the assumed distribution in samples of n, which are then ordered, $z_{(1)}, \ldots, z_{(n)}$. It will usually not take long to simulate 1000 such samples.

$$z_{1(1)}, \; z_{1(2)}, \qquad \cdots , z_{1(n)}$$
$$z_{1000(1)}, \; z_{1000(2)}, \; \cdots \; z_{1000(n)}$$

The ordered values are then averaged over the 1000 samples to give

$$\bar{z}_{(1)}, \; \bar{z}_{(2)}, \; \cdots , \bar{z}_{(n)}$$

These are estimates of the expected values of the ordered $x_{(i)}$, they thus act as the rankits in the previous section. The plot of $x_{(i)}$ versus $\bar{z}_{(i)}$ will give approximately a straight line if the model is valid. Deviations from the straight line will indicate the nature of any inadequacies in the model. Table 11.4 shows two sets of such simulated rankits used to compare a set of data with both Exponential and Pareto distributions.

Table 11.4 Simulated rankits

(a) Data having mean 9.49
   Simulated rankits from exponential with mean 9.49
   Simulated rankits from Pareto with mean 10.49 with one subtracted from each observation. (See Tables 3.3, 3.4.)
   Simulations based on 1000 samples of 20.

Exponential rankits	Ordered data	Pareto rankits −1
0.46	1.61	0.16
0.95	1.97	0.23
1.50	2.20	0.30
2.03	2.40	0.39
2.60	2.50	0.48
3.22	2.71	0.59
3.88	2.92	0.71
4.64	3.43	0.85
5.41	4.35	1.02
6.33	4.89	1.23
7.26	7.06	1.79
9.48	10.39	2.19
10.89	11.05	2.74
12.59	11.00	3.50
14.47	16.02	4.62
16.81	19.31	6.48
20.18	20.25	9.76
25.18	22.19	22.76
34.97	37.21	114.15

Data is closer to the Exponential than to the Pareto Distribution

Before leaving this topic it should be noted that for small quantities of data, all the various types of plot discussed in these sections show considerable variation from one sample to another. Hence a plot may look unlikely but still be consistent with the model being studied. It is worthwhile to use plots, based on equal amounts of data simulated from the model, to explore the range of possible variation in the plot.

11.2.4 Commonly available probability papers

Normal probability against linear, logarithmic, reciprocal
Normal probability against normal probability
Binomial
Poisson
Weibull
Extreme value
Chi-squared, with small degrees of freedom

Sources: UK: Chartwell.
USA: TEAM, Lowell, Mass.

11.3 Plots of fitted values

For regression and similar models a natural way to compare data with the model is to compare the observed values, y, of dependent variables with the corresponding fitted values, $\hat{y}$, obtained using the fitted model. A plot of y against $\hat{y}$, a *Fit-observation Diagram*, will give a series of points on a line, $y = \hat{y}$, at $45°$ through the origin, if the fit is perfect. This line we call the line of perfect fit. With any satisfactory model we would expect points close to and randomly scattered about this line. Any systematic deviation from the line indicates a lack of fit of the model used and gives some guide as to its nature. For example Figure 11.7 (a) shows a fit that is not very close but which might be acceptable. Figure 11.7(b) shows a systematic deviation with the fitted values, being too small for low and high values of $\hat{y}$ and too large for intermediate values. To examine the fit in relation to the effects of regressor variables it is often useful to plot both y and $\hat{y}$ against a regressor variable, x. Figure 11.7(c) shows such a plot. As usual we would expect no structure in the plot if the model is satisfactory. In the example shown the fitted values are smaller than the observations for small x and larger for large x. Because of the way $\hat{y}$ varies with x this systematic deviation would not show up without plotting against x. The fit-observation diagram suggests a natural numerical measure of quality of fit, since the correlation coefficient, $r_{y\hat{y}}$, between y and $\hat{y}$ will be $+1$ for a perfect fit and about zero for a plot showing no relationship at all. The square of this

$$R^2 = r^2{}_{y\hat{y}} = \frac{c^2{}_{y\hat{y}}}{s'_y{}^2 s'_{\hat{y}}{}^2}$$

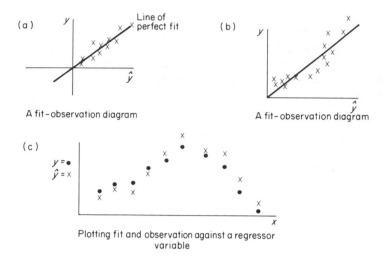

Figure 11.7 Fig–Observation plots

lies between 0 and 1 and is called the *Multiple Correlation Coefficient.* A value of R close to one suggests a well-fitting model.

11.4 Residual analysis

The standard form of many models is

$$y_i = \text{something} + e_i$$

where e_i represent independent random variables with zero mean, constant variance, and, possibly, a normal distribution. Having fitted the model we can replace the 'something' by $\hat{y}_i$, the appropriate fitted value. We can then write

$$y_i = \hat{y}_i + \hat{e}_i$$

If we have a good fit the $\hat{y}_i$ will be close to the unknown something and the quantities $\hat{e}_i$, the residuals, will be close to e_i. A well-fitted model will give residuals that are approximately independent random variables with zero mean, constant variance, and, possibly, a normal distribution. Table 13.1 gives the residuals for the road data of Chapter 1. If the residuals do not show the properties described then either the e_i do not have the properties assumed or the model is not the model we thought it was. Either way our model is shown to be inadequate and usually plots indicate where the inadequacies lie. There are many helpful ways of plotting the residuals, the following being the most common.

(a) Residuals against the observation order, i. Figure 11.8. This plot will show up any systematic changes occurring as the data collection

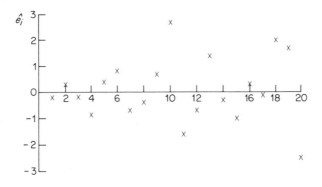

Figure 11.8 A plot of residuals in observation order

proceeded. For example measuring equipment may get out of adjustment, an experimenter may get tired and more uncertain in his
measurements, or allow the result of one observation to influence the
next observation.

(b) Residuals against the fitted value, $\hat{y}_i$. Figure 11.9. This plot will
indicate any systematic deviation of the underlying structure from the
model when this deviation varies with the magnitude of $\hat{y}$. It will also
indicate, as in the example shown, any relation between variance and
magnitude of $\hat{y}$. If such a lack of constant variance shows, it is
advisable to refit the model using a method suitable for non-constant
variance.

(c) Residuals against the observed values, y_i. Figure 11.10. This plot is
similar in interpretation to (b), it has the advantage of allowing more
random features to appear, in that y_i includes the effect of ε_i more
strongly than $\hat{y}_i$. However, it compares the residuals with the real
situation and not the assumed model which could be important if the
assumed model is not a close fit.

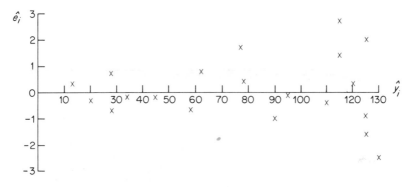

Figure 11.9 A plot of residuals against fitted valves

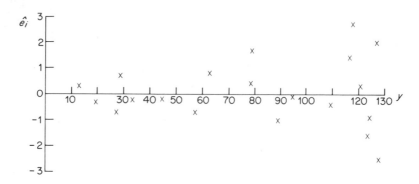

Figure 11.10 A plot of residuals against observations

(d) Residuals against regressor variables or new variables, x_i. Figures 11.11 and 11.12.

These plots are similar in objective to that illustrated in Figure 11.7(c). If the model is wrong about the way x_i influences y_i, then the error produced will be contained in the residual. A plot of residual against regressor variable will show a systematic variation in this situation. Figure 11.11 shows an extreme case of such a systematic variation with very little randomness. As x is already in the model the implication of this plot is that we have the variable in the wrong form. For example, if x is there only in linear form, this set of residuals suggests that terms up to x^3 might be included in the model. This form of plot is also useful for considering the inclusion of an entirely new variable. If x is not included in the current model, but is actually influencing y, then the residuals will include the influence of x. A plot of residuals against x

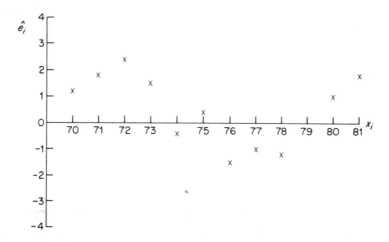

Figure 11.11 A plot of residuals against one regressor variable

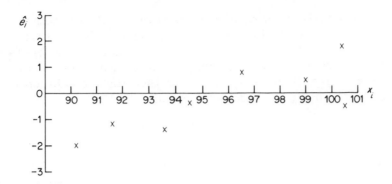

Figure 11.12 A plot of residuals against a new variable

will show this influence if it is at all important. This requires that we can find values x_i that correspond to the observations y_i. Figure 11.12 illustrates the situation by plotting the residuals from data simulated by a model of the form

$$y_i = \alpha z_i + \beta x_i + \varepsilon_i$$

when only

$$y_i = \alpha z_i + \varepsilon_i$$

was fitted. It will be seen that the linear term in x_i appears in the residuals.

The above plots consider the influence of only one variable. We may often be interested in the joint effect of several variables. Figure 11.13 shows the influence on the residuals of two variables z and x. Without using a three-dimensional presentation we can simply indicate the sign

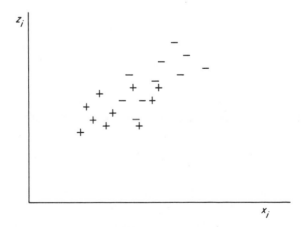

Figure 11.13 Residual sign against two regressor variables

of the residuals. In the example large or small values of both variables influence the sign of the residuals. This suggests that, if it is not there already, some term such as $z_i x_i$ should be tried in the model. If such a term is there then perhaps higher powers in products of u and x should be considered. If there is much data then the presentation may be clarified by using standardized residuals

$$\frac{\hat{e}_i}{s_{\hat{e}}},$$

where $s_{\hat{e}}$ is the standard deviation of the residuals, and plotting only standardized residuals of magnitude greater than some suitable value, two for example.

(e) Residuals against model components. Figure 11.14. Models are sometimes composed of a number of components each of which may include several variables or terms. In this case we may investigate a component in the same way as we investigated a single regressor variable. For example, a time series model may contain a seasonal component, S_i composed of a number of oscillatory terms. In the fitted model the seasonal term might be

$$S_i = 2.1 \sin \frac{2\pi i}{12} + 1.3 \cos \frac{2\pi i}{12} + 0.8 \sin \frac{4\pi i}{12} + 0.6 \cos \frac{4\pi i}{12}$$

Thus an investigation of the effectiveness of the seasonal, monthly, component would include a plot of $\hat{e}_i$ against S_i. As each S_i is repeated each year the actual form of plot will be as in Figure 11.14. The message of the plot of Figure 11.14 is that, as the residuals at each S_i are

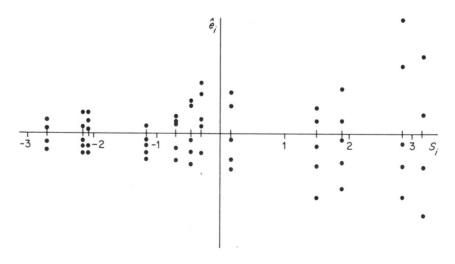

Figure 11.14 Residuals against seasonal component

scattered about zero, the seasonal component adequately models the seasonality in the data. However, it is clear that the random variation is not constant and that it increases with the seasonal component. If the current model includes the seasonal term plus random variation, then a modification of the model might be to use a multiplicative random variation, $S \times e$ instead of $S + e$. Such a multiplicative effect of randomness would explain the large residuals in the additive model when S was large.

(f) Residuals against residuals, Figure 11.15.

The usual model assumption of independence of the random variation, e_i, can be examined by a number of plots. A good check on this is given by a scatter diagram of paired residuals, with pairs being from adjacent observations in time or space or paired by some other suitable relation. Figure 11.15 shows such a plot which suggests a lack of independence. Where the data consists of a natural sequence of observations the correlations of residuals that are adjacent, two apart, three apart, etc. $(r_1, r_2, r_3, \ldots)$ can be calculated and the results plotted, the plot being called a *correlogram* (Figure 11.16). In this

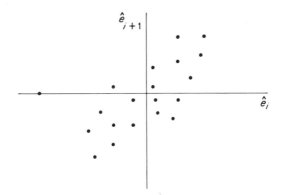

Figure 11.15 Plot of paired residuals

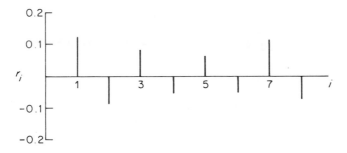

Figure11.16 A correlogram of residuals

144

example though the magnitudes of the autocorrelations are small, the alternating sign suggests a structure missing from the model.

(g) Residual distribution

The distribution of the residuals can be investigated using the plotting methods discussed in Section 11.2.

Chapter 12

Eclectic identification—formal methods*

12.1 Introduction

In the previous discussion we considered informal, exploratory, techniques to examine single models suggested by conceptual identification. It often occurs that conceptual identification leads to a number of possible models. The process of identification then becomes one of selecting between these alternatives. There are two particular situations that need separate consideration:

(a) In the most common case, conceptual and preliminary empirical modelling suggest that the model will be one within a family of models, for example a regression model or an autoregressive model. The problem then will be selection from that family, for example the selection of the regressor variables to include or the choice of the order of the autoregressive model. This approach is at its clearest in the analysis of Nested Models. The meaning of this term is best seen in the following two examples in which we use the notation of previous chapters.

$$y_i = \beta_0 + \beta_1 x_{1i} + e_i$$

is nested within

$$y_i = \beta_0 + \beta_1 x_{1i} + \beta_2 x_{2i} + e_i.$$

The model

$$x_t = \theta_1 x_{t-1} + e_t$$

is nested within

$$x_t = \theta_1 x_{t-1} + \theta_2 x_{t-2} + e_t.$$

Thus each model is a special case of the more general model in which some parameters are set at zero. Such families are called *Nested Models* or, sometimes, *Hierarchical Models*.

145

(b) The second situation is one where alternative models are available that are not nested. For example a lognormal distribution and a Gamma distribution might be possible alternatives.

The approaches to identification for nested and non-nested models are usually quite distinct, since in the nested case the ability to list and relate the possible models in the family greatly eases the analysis. In general only relatively small numbers of non-nested alternative models are considered, whereas very large numbers of possible models in a nested family are often compared.

12.2 Nested models

There are two extreme approaches to identification in nested models. The *Top-down* approach starts with the largest model of the family and proceeds by dropping terms. This largest model is sometimes called the *Saturated* model. If the saturated model fails to include a factor of real importance clearly the final model from the Top-down process will not include it. In consequence the saturated model tends to be very big, including everything that might conceivably be in the final model. A further consequence is that during the early stages of Top-down identification many parameters will require estimating, and hence the big models may not be very well fitted. This makes it difficult to distinguish between important and less important terms. In spite of these difficulties the Top-down approach has the great merit that, if the saturated model is correctly chosen, the best model available will be a subset of it, and proper statistical tests can be devised to successively eliminate terms not relevant.

The alternative approach is the *Bottom-up* where terms are brought into the model one at a time from a list of possible terms. At each stage the term that makes the greatest improvement is brought in to produce the next model. The process is then repeated. The Bottom-up approach is a consequence of the principle of simplicity. It seeks for the simplest model that will do the required job. If this approach is adopted then the fact that initially simple models are involved means that a relatively small number of alternatives need considering. This enables the use of exploratory techniques such as residual analysis and the incorporation of the modeller's prior information. The Top-down approach involves in its early stages so many possibilities that it tends to involve only automatic selection procedures. Both Top-down and Bottom-up procedures are stopped when further stages make no significant improvement to the fit of the model. The major problem with the standard Bottom-up approach, is that in the early stages the model being tried will be incomplete, for example important regressor variables will not yet have been included in the model. This means that the early fitted models may be seriously distorted and hence decisions based on them may be misleading.

The objections to both Top-down and Bottom-up may be partly met by a very thorough preliminary conceptual and empirical analysis. Then instead of starting with a saturated or single term model one starts somewhere near, but preferably above, a likely realistic model. The procedure then involves both the elimination and introduction of terms by using both Top-down and Bottom-up techniques. Thus a suitable mixture of the two approaches should normally be considered.

In the next section we illustrate the identification of nested models by considering regression models and the selection of regressor variables.

12.3 Selecting variables for regression models

The first step in the development of a regression model is to create a list of potential regressor variables, $X_0, \ldots, X_q$. If we consider the model that contains all of these variables, it is evident that each of its $q+1$ terms may either be included in or left out of any final model. There are thus $2^{q+1} - 1$ potential models that could be obtained from the list of potential regressors. The -1 removes from our consideration the 'non-model' in which all variables are left out.

It is instructive to consider some values of $2^{q+1} - 1$:

$$q = 3 \quad 2^{q+1} - 1 \quad = 15$$
$$q = 9 \quad 2^{q+1} - 1 \quad = 1023$$
$$q = 24 \quad 2^{q+1} - 1 \quad = 33,554,431$$

It is clear from these values that if q is small we could sensibly consider a range of properties of all the potential models. For $q = 9$ the prospect of doing a residual analysis on the 1023 models is daunting. If, however, we focussed on just one measure of the quality of the fitted model then, using a large and fast computer and some of the highly efficient formulae that are available for fitting regression models, it would be possible to compare the 1023 models. For $q = 25$ even this possibility is ruled out and we have to admit that we cannot consider all possible models. In this case we shall have to search for good models in some logical fashion, but it will have to be accepted that we can only look at a subset of the 33 million models and this subset may not contain the best models. In the following three subsections we consider briefly these three possible situations.

12.3.1 Small numbers of variables

We have in effect already dealt with the case of a small number of regressors. There are a few alternative models and it is feasible to carry out a residual analysis for each model as well as calculate a range of measures of quality of fit. The choice of 'best' model can thus be based on a broad consideration of relative merits. We may also carry out simple formal tests based on the residual sum of squares. Thus consider an initial regression model containing all the reasonable regressor variables. Suppose there are p terms in this model and we fit using n observations. The residuals, $\hat{e}_{i,p}$, from this fitted model can be used to produce the measure, s_p, of the variability of the observations about this fitted model. This is

$$s_p = \sum_{i=1}^{n} \hat{e}_{i,p}^2.$$

Suppose we now wish to test whether a particular group of m of the variables should be kept in the model. The residuals, $\hat{e}_{i,m}$, from this nested model give

$$s_m = \sum_{i=1}^{n} \hat{e}_{i,m}^2$$

The fact that the model underlying s_p contains more parameters than that giving s_m means it will give a better fit, so $s_p < s_m$. However, if the group m is a good subset and those terms dropped are of little consequence then s_m will be only a little larger than s_p. A measure of the quality of description provided by the group of m variables is

$$F = \frac{(s_m - s_p)/(p - m)}{s_p/(n - p)}$$

The divisors are included since it can be shown that F is distributed as $F_{p-m, n-p}$ provided the removed parameters, $\beta_{m+1}, \ldots \beta_p$, are all zero. An F-test can thus be carried out to test the quality of the nested model. If F is significantly large it indicates that we have dropped one or more significant variables.

If we carry out a small number of these tests we can interpret the significance levels in the proper manner of significance testing. In dealing with medium and large numbers of regressor variables, tests of this form are sometimes used, but they no longer have a clear interpretation in terms of significance levels. Such a 'test' may be made hundreds of times and in almost all cases both models being compared will be invalid. Thus the test statistics are simply being used as indicators to guide in a search procedure.

12.3.2 Medium numbers of variables

It has been suggested that to compare a hundred or possibly even a thousand possible models we need simple summary measures of quality of fit. The most natural measure is the residual sum of squares. Table 12.1 gives the residual sum of squares for all possible models for a situation with only four possible regressor variables. Obviously the best fit is obtained by including all four variables. Dropping T or W does not lead to a much worse fit, but dropping M or P makes the fit much worse. This suggests that M and P are the important terms, a conclusion confirmed by

Table 12.1 Listing all possible residual sums of squares. Variables M, T, W, P.

Variables included	Residual S S	Degrees of freedom
$M\ T\ W\ P$	5383	14
$-T\ W\ P$	8767	15
$M-W\ P$	5408	15
$M\ T-P$	5439	15
$M\ T\ W-$	8983	15
$--W\ P$	8795	16
$-T-P$	9182	16
$-T\ W-$	11710	16
$M---P$	5440	16
$M-W-$	8983	16
$M\ T--$	8986	16
$M---$	8992	17
$-T--$	11830	17
$--W-$	11720	17
$---P$	9241	17
$----$	11860	18

the small residual sum of squares for the model involving only M and P. A problem that arises here is that adding more terms to a model naturally leads to a better fit. Thus the models with M, P, T and M, P, W both have smaller residual sums of squares than that involving M and P alone. However, the improvement of fit is small, so the principle of parsimony suggests that we keep to the M,P model. This intuitive allowance for the number of parameters is feasible for a four-variable example; we can look sensibly at 15 residual sums of squares. With a thousand this is much more difficult and it would be much better if we could adjust in some way for the number of parameters. This would enable a computer search for the best fitting models, by simply seeking the smallest (or highest) value of a suitable criterion function. A number of criteria have been used of which we shall mention three.

(a) The Adjusted Multiple Correlation, $\overline{R}_k^2$.

The ordinary multiple correlation coefficient, R^2, is the square of the correlation between the observations, y_i, and the fitted observations, $\hat{y}_i$. If there are k parameters in the model and n observations, $y_1, \ldots, y_n$, then the larger k the better the fit and the greater the value of R^2. The following adjustment for k leads to a reduction as k increases:

$$\overline{R}_k^2 = 1 - (1 - R^2)\frac{n}{n-k}$$

Though R^2 automatically increases as a new parameter is introduced, $\overline{R}^2$ does not and we can compare $\overline{R}^2$ for models containing different numbers of parameters. The models giving the highest $\bar{R}^2$, closest to one, are taken as the best fitting.

(b) Mallows C_p.

This is a direct adjustment of the residual sum of squares (RSS) based on p variables in the model. It takes the form:

$$C_p = \frac{RSS}{\hat{\sigma}^2} + 2p - n,$$

where $\hat{\sigma}^2$ is an estimate of the underlying variance σ^2, usually based on fitting all the regressors. The criterion C_p does somewhat more than adjust RSS, for if a correct model is used $E(C_p) = p$. Thus large deviation of C_p from the line $C_p = p$ suggests a wrong model and consequent bias in the fitted model.

(c) Prediction Sum of Squares (PRESS)

Though a model's fit improves with the number of variables included, its predictive ability for new data does not necessarily improve. Given new, validation, data $y_{n+1}, \ldots, y_{n+v}$, with the corresponding regressor variable data, then a measure of model quality is obtained by using the model to obtain predictors $\hat{y}_{n+1}, \ldots, \hat{y}_{n+v}$. The natural criterion is the predictive sum of squares

$$PRESS = \sum_{i=1}^{v} (y_{n+i} - \hat{y}_{n+i})^2$$

The set of regressors leading to the smallest PRESS gives the best model. Often such validating data does not exist so we make use of what can be called *Jackknife Validation*. We predict y_i, within the data $y_1, \ldots, y_n$, by using the fitted model based on all the observations except the ith, y_i, giving $\hat{y}_{(i)}$. For any set of regressors we can do this for all n observations separately and obtain

$$PRESS = \sum_{i=1}^{n} (y_i - \hat{y}_{(i)})^2.$$

In each term $(y_i - \hat{y}_{(i)})^2$ we use a prediction of y_i based on 'other data'. PRESS will not necessarily decrease just by increasing the number of regressors, it gives a genuine measure of the quality of the model being used. Clearly the evaluation of PRESS for many variables and many observations involves a great deal of computation but it is a natural and effect criterion worthy of more extensive use.

12.3.3 Large numbers of variables

As discussed in section 12.2 there are two extreme approaches to handling a large number of regressors. The Top-down approach is to start with all variables in the model and then systematically eliminate variables according to some criteria. In regression this is usually called *Backward Elimination*. The Bottom-up approach, called *Forward Selection*, starts with nothing and builds up the model by adding variables. A natural extension is to synthesize these two approaches; such approaches are called *Stepwise Methods*. Suppose we have at some stage in any of these procedures a model containing p variables out of a total set of q variables. Then what we require are procedures for putting in order of likely importnce (a) the p variables in the current set, and (b) the $q - p$ variables not in this set. There are a number of ways of producing such ordered lists appropriate in different circumstances. We give three only.

(a) An ordered list may be based on conceptual considerations. Most computer methods ignore such possibilities but there is much merit in allowing such conceptual orderings to influence what would otherwise be purely empirical rules. Sometimes variables might be included in the model irrepective of computer results.

(b) If we estimate the coefficients, β_i, for the p variables in the model, standardize by dividing by the estimate's standard error we get

$$t_i = \hat{\beta}_i / s_{\beta_i} \qquad i = 1, \ldots, p.$$

Table 12.2(a) illustrates a four-parameter model. If we wish to test the hypothesis that a particular $\beta_i = 0$ then t_i would be a possible test statistic. Thus if $|t_i|$ is small we could take it that $\beta_i = 0$ and leave out variable x_i. The probable lack of independence of the estimators makes formal testing hard to interpret but the $|t_i|$ provide a natural ordering for the variables. In the example of Table 12.2, T has the smallest $|t_i|$ and this would be the natural variable to exclude in moving to a three-variable model.

(c) We have already referred to the changes in the residual sum of squares produced by adding variables to a model. A natural procedure is based on examining the magnitude of these changes when at any stage variables are added or subtracted. Table 12.2(b) illustrates such changes. In dropping variables it is reasonable to drop those that produce the least improvement and add those that produce the most improvement. These are starred, *, in the table.

In backward elimination we may use any of the above approaches to move from p to $p-1$ variables, starting with $p = q =$ all the variables. For forward selection (c) provides a natural approach, though (b) can be modified by using t values for non-included variables. In stepwise regression the procedure is to seek to follow backward elimination, but having moved from p to $p-1$ an examination is made of all the variables left out to see if a significant improvement can be obtained by moving back to p with a new variable being added.

The above procedures provide ways of adding new variables or deleting existing variables at any stage. To complete a search technique there is a need for a *Stopping Rule* that stops the search at a particular model as the

Table 12.2

(a) Parameter for constant	Estimate $\hat{\beta}$	Standard error	$|\hat{\beta}/s_\beta|$
M	-103.7	48.0	2.16
T	0.92	3.7	0.25
W	0.11	0.43	0.26
P	1.19	0.39	3.05

(b) Changes in Residual Sum of Squares
(i) Top down
From four-variable model by dropping:

M	3384	
T	25*	$\rightarrow$ suggests M + W + P
W	56	
P	3600	

From three-variable model $M + W + P$ by dropping

M	3387	
W	32*	$\rightarrow$ suggests M + P
P	3575	

(ii) Bottom up
From one-variable model M by adding

T	6	
W	9	
P	3552*	$\rightarrow$ suggests M + P

* Best candidate for dropping or adding.

'best'. In essence such stopping rules examine the values of $|t|$, or of the change in the residual sum of squares. We would either (i) not remove the variable identified since its $|t|$ is too large or (ii) not bring in the variable identified, if for example it only produces a negligible reduction in the sum of squared residuals.

There are many approaches to variable selection and different computer packages implement different rules.

12.4 Bayesian methods

Bayesian Methods make use of prior probabilities assigned to models as discussed in section 8.8. The formula that is used is called *Bayes Formula* after the originator of this approach, the Rev. Thomas Bayes (1763), and the methodology is referred to as *Bayesian Methods*. The formula is concerned with the reversal of conditional probabilities. If we have two related events A and D, $Pr(A|D)$ is the probability of A occurring given, conditional on, D having occurred. The multiplication law of probabilities

in this situation takes the form
$$Pr(A \text{ and } D) = Pr(A|D)Pr(D).$$
But we could also write this
$$Pr(A \text{ and } D) = Pr(D|A)Pr(A)$$
and hence have
$$Pr(A|D)Pr(D) = Pr(D|A)Pr(A).$$
This formula enables us to reverse conditional probabilities from $Pr(D|A)$ to $Pr(A|D)$. Suppose now that A represents the occurrence of either of two models M_0 or M_1 with probabilities.
$$Pr(A = M_0) = Pr(M_0), \quad Pr(A = M_1) = Pr(M_1)$$
which are the *Prior Probabilities* of these models. Given either model we can evaluate the probability, $Pr(D|M_i)$, $i = 1$, or 2, of obtaining our data, which we shall denote by D and for the moment assume to be from a discrete distribution. The previous formula can now be used to reverse the conditional probabilities giving
$$Pr(M_0|D) = \frac{Pr(D|M_0)Pr(M_0)}{Pr(D)}$$
$$Pr(M_1|D) = \frac{Pr(D|M_1)Pr(M_1)}{Pr(D)}$$
The probabilities $Pr(M_i|D)$ are called *Posterior Probabilities* and represent the evidence for choosing between M_0 and M_1, allowing for both our prior probabilities and our new data. To calculate the probabilities $Pr(M_i|D)$ we need to be able to obtain the unconditional probability, $Pr(D)$, of the data. This is obtained by summing out over the possible models
$$Pr(D) = Pr(D \text{ and } M_0) + Pr(D \text{ and } M_1)$$
which we could also express as
$$Pr(D) = Pr(D|M_0)Pr(M_0) + Pr(D|M_1)Pr(M_1).$$
Usually we are not interested in the actual posterior probabilities of the two models but only their relative sizes. This is expressed as the *Odds*. The *Prior Odds* being $Pr(M_1)/Pr(M_0)$, the *Posterior Odds* being $Pr(M_1|D)/Pr(M_0|D)$. Using the previous formulae it is seen that
$$\frac{Pr(M_1|D)}{Pr(M_0|D)} = \frac{Pr(D|M_1)}{Pr(D|M_0)} \cdot \frac{Pr(M_1)}{Pr(M_0)}.$$
The first term on the right-hand side is the ratio of the probabilities of the data given M_1 and M_0. If the data comes from continuous distributions then the probabilities are replaced by what in section 2.6 we termed Initial Likelihoods, $\mathscr{L}^*(M)$. For discrete data the two quantities are identical, thus in general we have
$$\frac{Pr(M_1|D)}{Pr(M_0|D)} = \frac{\mathscr{L}^*(M_1)}{\mathscr{L}^*(M_0)} \cdot \frac{Pr(M_1)}{Pr(M_0)}.$$

Thus the odds between the two alternative models given the data is obtained from the prior odds and the ratio of Initial Likelihoods. We will illustrate the use of this formula in the following sections. For the moment it is sufficient to make the following observations.

(a) The formula provides a general means of bringing together the conceptual evidence, in the form of prior probabilities, and empirical evidence, in the form of likelihoods. It consequently focusses on prior probabilities and likelihoods as the means by which evidence needs to be formulated for this methodology to apply.
(b) The method requires that $Pr(M_0) + Pr(M_1) = 1$ so that these two models are assumed to be the only possible alternatives.
(c) If we have no prior evidence favouring one or other of the models, then prior probabilities of 0.5 would be appropriate giving prior odds of one. In this case the ratio of the likelihoods will provide the posterior odds.
(d) A final point, that we will illustrate later, is that we need the likelihoods of the models without reference to the model parameters. The process of removing the influence of the model parameters can be computationally very difficult.

12.5 General methods of choice for nested models

Section 12.3 illustrated the techniques for seeking the best regression model without indicating the formal basis for the techniques used. The objective of this section is to describe briefly some of the methodology underlying these techniques and to indicate their applicability to a much wider range of models than simply regression models. The discussion of regression has emphasized the role of the residual sum of squares as the measure of the quality of fit. For linear models with normally distributed errors this is the natural measure to use and it is in fact proportional to the log likelihood. When we move to other models and distributions we start with the idea of the likelihood function, or its logarithm, as providing a more general measure of quality of fit.

Example 1

Consider a standard regression model

$$y_i = \mu_i + e_i, \quad e_i \cap N(0, \sigma^2)$$
$$\mu_i = \Sigma \beta_j x_{ij}$$

The normal distribution for e_i gives

$$f(e_i) = \frac{1}{\sqrt{2\pi}\,\sigma} \exp -\frac{1}{2\sigma^2}(e_i)^2$$

For the n observations this gives the initial likelihood

$$\mathscr{L}^*(\beta) = \prod_{i=1}^{n} f(e_i),$$

$$\mathscr{L}^*(\beta) = \left(\frac{1}{\sqrt{2\pi}\,\sigma}\right)^n \exp -\frac{1}{2\sigma^2}\sum(y_i - \mu_i)^2$$

If we replace the β by their estimates and note that the residuals are defined by

$$\hat{c}_i = y_i - \hat{\mu}_i, \text{ where } \hat{\mu}_i = \sum_j \hat{\beta}_j r_{ij}, \text{ then}$$

$$\mathscr{L}^*(\hat{\beta}) = \left(\frac{1}{\sqrt{2\pi}\,\sigma}\right)^n \exp -\frac{1}{2\sigma^2}\sum\hat{e}^2$$

The log likelihood, neglecting constant terms, is thus

$$L(\hat{\beta}) = -\frac{1}{2\sigma^2}\sum\hat{e}^2$$

which is simply a negative multiple of the residual sum of squares. Thus finding the $\hat{\beta}$ for the model with the smallest residual sum of squares is equivalent to finding the $\hat{\beta}$ for the model with the largest log likelihood or likelihood.

The logic of the comparisons of this last example requires that the choice of estimators, $\hat{\beta}$, used in the fitted model should be those that give maximum values to the likelihood of the model under consideration. This will be discussed further in the chapters on estimation. For now it is sufficient to argue that to compare two nested models we need simply compare the largest values that their respective likelihoods can attain. This is most easily done by looking at the ratio

$$\lambda = \frac{\mathscr{L}(\hat{\beta}_0|M_0)}{\mathscr{L}(\hat{\beta}_1|M_1)}$$

where M_0 and M_1 are the two models to be compared and the $\hat{\beta}$ are the best estimates, in the sense of maximizing the likelihood of the parameters of the model. With nested models any multiplying terms not involving the parameters will cancel out. If λ is much greater than one M_0 is the preferred model, if much less than one M_1 is the preferred model.

Example 2
Consider a nested regression situation where model M_1 has parameters β_0, β_1, β_2 and model M_0 is the next model down with $\beta_2 = 0$. If we fit both models and obtain their residual sums of squares we get

$$\text{RSS}_1 = \sum_i \hat{e}_{1i}^2$$

and

$$\text{RSS}_0 = \sum_i \hat{e}_{0i}^2.$$

Assuming normality the likelihood ratio λ is

$$\lambda = \exp - \frac{1}{2\sigma^2} \left(\sum_i \hat{e}_{0i}^2 - \sum_i \hat{e}_{1i}^2 \right).$$

Thus if σ^2 is known a test of M_0 versus M_1 can be based on $\text{RSS}_0 - \text{RSS}_1$, if this is large and positive M_1 is preferred, if large and negative M_0 is selected.

This intuitive rule can be turned into a formal test, independently of knowledge of σ, by using the fact that

$$\frac{(\text{RSS}_0 - \text{RSS}_1)}{\text{RSS}_1 /(n-3)} \cap F_{1,n-3}$$

so an F test can be used to test the hypothesis that $\beta_2 = 0$.

The likelihood ratio in the above example required a transformation to obtain a test statistic with a standard distribution. For large samples a useful approximate distribution can be expressed as

$$-2 \ln \lambda \cap \chi^2.$$

The degrees of freedom of the χ^2 is the number of parameters set to zero in the nesting of M_0 within M_1. Using log likelihoods we have

$$-2 \left[L(\hat{\beta}_0 | M_0) - L(\hat{\beta}_1 | M_1) \right] \cap \chi^2$$

The basic quantity in approximate tests of M_0 and M_1 is thus $-2L(\hat{\beta}|M)$ which is called the *Deviance*, $D(M)$.

Example 3
Consider a situation where the mean number of faults per metre of wire, λ, produced on a machine was thought to depend on a temperature variable x and an internal tension variable z. The model for λ used was

$$\lambda = e^{-(\alpha + \beta x + \gamma z)}$$

For fixed x_i, z_i the distribution of the length y_i between faults is exponential Expn (λ_i). Thus the likelihood based on independent observations (x_i, z_i, y_i) is

$$\mathscr{L}(\alpha, \beta, \gamma) = (\Pi \lambda_i) e^{-\Sigma \lambda_i y_i}$$

and the log likelihood is

$$L(\alpha, \beta, \gamma) = -\Sigma(\alpha + \beta x_i + \gamma z_i) - \Sigma \{ y_i e^{-(\alpha + \beta x_i + \gamma z_i)} \}$$

To test whether z_i makes a meaningful contribution to the model we find values $\hat{\alpha}$, $\hat{\beta}$, $\hat{\gamma}$ to give the maximum log likelihood for the complete model giving a deviance

$$D(M_1) = -2L(\hat{\alpha}, \hat{\beta}, \hat{\gamma}).$$

Similarly for the nested model with $\gamma = 0$ values $\hat{\alpha}, \hat{\beta}$ are found to give the maximum $L(\alpha, \beta, 0)$ and hence

$$D(M_0) = -2L(\hat{\hat{\alpha}}, \hat{\hat{\beta}}, 0).$$

The numerical values of such quantities are provided by computer packages such as GLIM. If $\gamma = 0$ then $D(M_0) - D(M_1)$ will have approximately a χ_1^2 distribution and so can be used to test the hypothesis that $\gamma = 0$.

12.6 Choosing between non-nested models

In the current state of the craft the structure in nested models enables the use of search procedures to compare many different models. The lack of such structure for non-nested models means that most methods are designed for choosing between only a few alternatives, often just two. The informal techniques of residual analysis and probability plotting provide an obvious first step in comparing small numbers of models. In this section we shall discuss three more formal approaches, those based on Bayes formula, on measures of quality of the model and on seeking to relate the models.

12.6.1 Bayes choice

Suppose that we have alternative models $M_i\ i = 1, 2, \ldots, k$. for which we have prior probabilities, $P(M_i)$, which cover all possibilities so that $\sum_i P(M_i) = 1$. If the observations are represented by the symbol D we may write, by the formulae of section 12.4,

$$Pr(M_i|D) = Pr(D|M_i)Pr(M_i)/Pr(D)$$

where

$$Pr(D) = \sum_i Pr(D \text{ and } M_i)$$

$$= \sum_i Pr(D|M_i)Pr(M_i).$$

This gives the posterior probability associated with model M_i given the data D. The most appropriate model will be the one with the largest posterior probability. It is possible to extend this methodology to examine the costs associated with wrong choices but we will leave that to be dealt with later in the Application Section. It should be noticed that the right-hand term $Pr(D|M_i)$ is, in the language of section 2.6, the initial likelihood of model M_i, $\mathscr{L}^*(M_i)$. The formula can thus be equivalently written

$$Pr(M_i|D) = \mathscr{L}^*(M_i)Pr(M_i)/Pr(D)$$

The fact that we are comparing initial likelihoods from totally different models requires that we must not drop multiplying constants from the likelihood as is often done, since in this application they do not cancel out in the calculations. If we wish to compare only two models the formula can be used to give *Posterior Odds*

$$\frac{Pr(M_0|D)}{Pr(M_1|D)} = \frac{\mathscr{L}^*(M_0)}{\mathscr{L}^*(M_1)} \cdot \frac{Pr(M_0)}{Pr(M_1)},$$

in words

Posterior Odds = Initial Likelihood Ratio × Prior Odds.

Taking logs of the relation gives an additive form:

Posterior log odds = log initial likelihood ratio + Prior log odds

It will be seen that all the evidence provided by the data as to the choice of model is condensed into the Initial Likelihood Ratio. If we have $50:50$ odds for M_0 and M_1 the prior log odds is zero, so the posterior log odds is equal to the log initial likelihood ratio. The procedure suggested by these results is that we should use the model giving the largest initial likelihood. The initial likelihoods will depend not only on the models but also on the unknown parameters in the models. We should thus really write $\mathscr{L}^*(M_0, \theta_0)$. It is evident that to compare M_0 and M_1 we need to remove the effect of the parameters. There are two ways of doing this.

(a) The first method is to replace θ by some numerical estimate $\hat{\theta}$. As we are looking for the largest $\mathscr{L}^*(M_i, \theta_i)$ it is reasonable to use the value of $\hat{\theta}$ that maximizes $\mathscr{L}^*(M, \theta)$. Such quantities, called *Maximum Likelihood Estimators*, can usually be easily derived from the data. We discuss their derivation in section 13.4. This approach though it started with Bayes formula is essentially an empirical and informal means of comparison. We choose equal prior probabilities for the competing models, as we wish only to consider the empirical evidence, and we use the data as the means of removing the parameters, θ. The choice of model is hence based on a numerical comparison of $\mathscr{L}^*(M_0, \hat{\theta}_0)$ and $\mathscr{L}^*(M_1, \hat{\theta}_1)$. Before illustrating this approach it should be noticed that for nested models the initial likelihoods become standard likelihoods and this discussion gives some justification for the comparison of maximized likelihoods which was used in the choice of nested models.

(b) An alternative approach makes full use of Bayesian methods by seeking prior distributions, $h(\ )$, for the parameters. The choice of model is then based on comparison of expected values

$$\mathcal{L}_E^*(M) = E_\theta \left[\mathcal{L}^*(M, \theta) \right].$$

The use of expectation with respect to the prior distribution removes the occurrence of the unwanted parameters, though it may introduce additional parameters associated with the prior distribution. The specification of appropriate priors and the calculation of $\mathcal{L}_E^*(M)$ is, for most distributions, rather involved and method (a) provides a simpler approach.

Example 1
A discrete random variable R can take values in the range $r = 0, 1, 2, \ldots, \infty$. It is decided to use either a Poisson distribution $(M_0 = P)$ or a slight variation on the Geometric distribution $(M_1 = G)$ as a simple approximate model. The Geometric model is put in the form

$$Pr(R = r) = (1 - \phi)\phi^r, \ r = 0, 1, 2, \ldots.$$

Thus, given n observations $r_1, \ldots, r_n$, the probabilities of the data (the initial likelihoods) are:

$$Pr(D|P) = e^{-n\mu} \frac{\mu^{\Sigma r_i}}{\prod r_i!}$$
$$Pr(D|G) = (1 - \phi)^n \phi^{\prod r_i}$$

To eliminate the effect of the unknown parameters they are replaced by their estimators, $\hat{\mu}$ and $\hat{\phi}$. It will be shown later that these take the form

$$\hat{\mu} = \Sigma r / n \qquad \text{and} \qquad \hat{\phi} = \Sigma r / (n + \Sigma r)$$

Substituting these gives the estimated probabilities as

$$P(D|P, \hat{\mu}) = e^{-\Sigma r} \frac{(\Sigma r)^{\Sigma r}}{n^{\Sigma r} \prod r!}$$
$$P(D|G, \hat{\phi}) = \left(\frac{n}{n + \Sigma r} \right)^n \left(\frac{\Sigma r}{n + \Sigma r} \right)^{\Sigma r}$$

The ratio of the estimated posterior probabilities simplifies to

$$\frac{Pr(P|D)}{Pr(G|D)} = \frac{e^{-\Sigma r}}{\prod r!} \cdot \left[\frac{n + \Sigma r}{n} \right]^{(n + \Sigma r)} \cdot \frac{Pr(P)}{Pr(G)}$$

Table 12.3 shows three simple sets of ten observations expressed in semigraphic form.

It will be seen that, assuming no prior preferences, as the data moves to a more symmetrical form, which is inconsistent with the geometric distribution, so the ratio of posterior probabilities moves from 1 to a substantial ratio, clearly pointing to the choice of a Poisson model.

12.6.2 Using criteria and tests

Some of the criteria used in comparing nested models still have relevance for non-nested models. For example the multiple correlation coefficient could be used to compare different models. A further criterion which has been used in the non-nested case is based on the likelihood function with best estimates of the parameters and an adjustment for the number of parameters. This criteria called the Akaike Information Criterion, AIC, is defined as

$$\text{AIC} = -2\ln \mathscr{L}(\hat{\theta}) + 2p$$

where $\hat{\theta}$ represents the estimators of the p parameters symbolized by θ. The term $2p$ adjusts for the number of parameters in the same way as it did in the Mallows C_p coefficient of section 12.3.2. Indeed for regression models with normal errors the AIC corresponds closely to C_p. The smaller the value of AIC the better the model.

Example 1 (continued)
Returning to the Poisson and Geometric comparison the AIC takes the form, for the Poisson, of

$$\text{AIC} = -2\ln P(D|P, \hat{\mu}) + 2$$
$$= 2\sum r(1 + \ln(n) - \ln(\sum r) + 2\ln(\prod r_i!) + 2$$

For the Geometric distribution it is

$$\text{AIC} = 2(n + \sum r)\ln(n + \sum r) - 2n\ln(n) - 2\sum r\ln(\sum r) + 2$$

Table 12.3 gives these values for the three sets of data and again it is clear

Table 12.3 Bayes choice of model

Prior probabilities—assume $Pr(P) = Pr(G) = 0.5$
Three sets of ten observations

	Data (a)	Data (b)	Data (c)
	0	1	1
	0 1	0 1	1 2
	0 1 2	0 1 2	0 1 2
	0 1 2 . 4	0 1 2 . 4	0 1 2 . 4
$n =$	10	10	10
$\sum r_i =$	11	12	14
$\prod r_i! =$	96	96	192
$\dfrac{Pr(P\|D)}{Pr(G\|D)} =$	1.0	2.2	42.7
AIC for $P =$	31.03	30.75	31.09
AIC for $G =$	31.06	32.32	34.60

that for data (a) there is nothing to choose between the two models but for the data of (b) and (c) the AIC is clearly indicating that the Poisson is better.

Notice that the AIC is a criterion of the quality of the model, it does not provide a formal test. The AIC is particularly useful if a large number of non-nested models are to be compared.

To find a formal test we first note that in studying nested models we made use of tests based on the log maximum likelihood ratio:

$$\ln \lambda = L(\hat{\beta}_0 | M_0) - \bar{L}(\beta_1 | M_1),$$

which had an approximate χ^2 distribution. With non-nested models, M_0 and M_1, the parameters β_0 and β_1 will be unrelated. Suppose we take M_0 as the model for the null hypothesis and denote the expectation of $\ln \lambda$ under this hypothesis by $E_{M_0}(\ln \lambda)$. A suitable test statistic is then

$$T = \ln \lambda - E_{M_0}(\ln \lambda),$$

due to Cox, see references. The statistic T is asymptotically $N(0, \sigma_T^2)$, under H_0 and provides a test of M_0 against the alternative M_1. The quantities $E_{M_0}(\ln \lambda)$ and σ_T^2 can be found by calculation in suitable cases, by approximation or by using simulation. If simulation is used the observed $\ln \lambda$ can be compared directly with simulated values generated using each of the two models.

12.6.3 Embedding

The basic idea of embedding is to devise a new model M_2 which has both the non-nested models M_0 and M_1 as special cases.

Example 1
In choosing between the Poisson and Geometric we can devise a '*mixed distribution*'

$$Pr(R = r) = p \cdot e^{-\mu} \frac{\mu^r}{r!} + (1-p)(1-\phi)\phi^r$$

The parameter p may then be estimated and a value close to 0 or 1 will identify models G and P respectively. The problem of identification is thus turned into one of estimation. An alternative approach is to note that if R is *either* Poisson or Geometric then $p = 0$ or 1. Therefore given observations $r_1, \ldots, r_n$ the likelihood is

$$\mathscr{L}(p, \mu, \phi) = \prod_1^n \left[pe^{-\mu} \frac{\mu^{r_i}}{r_i!} + (1-p)(1-\phi)\phi^{r_i} \right]$$

$$= p \prod_1^n e^{-\mu} \frac{\mu^{r_i}}{r_i!} + (1-p) \prod_1^n (1-\phi)\phi^{r_i}$$

since $p^n = p$, $(1-p)^n = (1-p)$ and $p(1-p) = 0$
so

$$\mathscr{L}(p, \mu, \phi) = p\mathscr{L}^*(\mu) + (1-p)\mathscr{L}^*(\phi)$$

where $\mathscr{L}^*(\mu)$ and $\mathscr{L}^*(\phi)$ are the initial likelihoods for the Poisson and Geometric distributions. Given prior probabilities π and $1 - \pi$ of these two distributions and appropriate priors for μ and ϕ then Bayes Formula can be applied to get the posterior probabilities of the two models as in Example 1.

12.7 Experiments for identification

Seeking to identify models using data from situations over which no control has been exercised can be a highly misleading exercise. The fact that two variables have a high correlation is no justification to assume any causal relation between them, the correlation may just reflect the fact that both variables are increasing with time. If, however, in a controlled experiment, changes in x are observed to produce changes in y then we are on much firmer ground. Often prior knowledge will suggest a relationship but not the detailed form of the model. These considerations suggest that, wherever possible, identification should be based on data from a designed experiment. Again we are touching on a large area of study and can only illustrate the subject.

Example 1
Consider the problem of modelling the relation between the output of a chemical process and the temperature at which it takes place. The natural experiment would be to measure the output, y_i, at a series of set temperatures, x_i, leading to typical regression data. If we have no inkling of the form of the model, then it makes intuitive sense to spread our set values of temperature evenly over as wide a range of temperatures as possible. The evenness of spacing gives us the best chance of seeing the structure. We would plot the data, try some natural models and carry out residual analyses to seek the model. Suppose, however, that on the basis of either prior knowledge or the principle of simplicity we decide only to consider two possible models:

$$y = \beta_0 + \beta_1 x + e$$

and

$$y = \beta_0 + \beta_1 x + \beta_2 x^2 + e.$$

In this case the objective of the experiment is to choose between a straight line and a curved line. Intuition, and theory, suggest that this distinction can best be seen by looking at the highest and lowest practicable and reasonable temperatures, x_h and x_l, and at the middle temperature $x_m = \frac{1}{2}(x_h + x_l)$. If we put a quarter of the observations at each of the end points and the rest at x_m we will have a more effective experiment for distinguishing between these two models. The essential point of this example is that consideration of prior knowledge is an essential ingredient in designing experiments for identifying models.

Example 2
As a second example consider the problem of identifying simple discrete linear systems. These occur in many areas of application and take the form

$$y_i = \sum_{r=0}^{i} c_{i-r} x_r \qquad \text{(see section 7.2.2).}$$

where $x_0, x_1, x_2, \ldots$ are the inputs to the system and $y_0, y_1, y_2, \ldots$ are the outputs produced. The identification of this linear system amounts to deciding how many of the coefficients, c, are non-zero and estimating what their values are, or possibly fitting some simple mathematical function to the sequence $\{c_j\}$. A deterministic way of finding the weights c_j is to give the system a kick and watch its response. Putting this rather more formally we make the input an impulse function

$$x_0 = 1$$
$$x_i = 0 \quad i = 1, 2, \ldots.$$

It will be seen from the linear model that with this input the output $y_i = c_i$. Thus the output exactly reflects the weights of the linear system, in fact the weights c_i are sometimes referred to as the impulse response function for the system.

If the system is a chemical plant or an aircraft in flight, whose characteristics can change quite significantly in different operating conditions, then stopping everything and giving the system a kick is hardly a practicable proposition. In these circumstances a statistical approach can be adopted that adds to the existing inputs an additional random signal, $\{e_i\}$. As the system is linear, the output of the system will be altered by an amount

$$v_i = \sum_{r=0}^{i} c_{i-r} e_r.$$

This can be re-written as

$$v_i = \sum_{r=0}^{i} c_r e_{i-r}.$$

We assume that $c_r = 0$ for negative r, which simply implies that v_i cannot be influenced by future values of e. Suppose the random signal has expectation zero and some covariance $\text{Cov}_e(e_i, e_{i+s})$ which we will write as $\gamma_e(s)$ for a stationary signal. Consider the covariance, $\gamma_{ve}(k)$, between the output v at time $i+k$ and the input e at time i.

$$\text{Cov}\,(v_{i+k}, e_i) = E(v_{i+k}e_i)$$

$$= E\left[\sum_{r=0}^{i+k} c_r e_{i+k-r}e_i\right]$$

$$= \sum_{r=0}^{i+k} c_r E(e_{i+k-r}e_i)$$

so

$$\text{Cov}\,(v_{i+k}, e_i) = \sum_{r=0}^{i+k} c_r \gamma_e(k-r)$$

It will be seen that the right-hand side depends on i so the process being dealt with is not formally stationary. This, however, is essentially due to the effects of 'switching on' the system. If the random signal has a covariance that dies away for large lags and the system has been operating for some time, so that i is large, the relationship can be replaced by

$$\gamma_{ve}(k) = \sum_{r=0}^{\infty} c_r \gamma_e(k-r).$$

Thus the covariance between input and output, the cross-covariance of lag k, is equal to the output of the system that is obtained if the input is the autocovariance of $\{e_i\}$. The autocovariance that corresponds to the impulse function of the deterministic method of identification is that for a white noise signal. That is, the input signal is a series of independent random numbers with zero mean and variance σ^2. Hence

$$\gamma_e(0) = \sigma^2$$

$$\gamma_e(s) = 0 \quad s = 1, 2, 3, \ldots$$

and so

$$\gamma_{ve}(k) = \sigma^2 c_k.$$

Hence the impulse response function c_k of the system can be determined by direct measurement of the autocovariance, $\gamma_{ve}(k)$, when an independent random signal is used as input. The input random sequence is designed so that (i) it can be simply superimposed on the operating inputs of the system and (ii) the response $\{v_i\}$ can be clearly distinguished from the operating outputs. Often use is made of *Binary Random Sequences*, e.g. 00101110 . . . where the same random sequence is repeated to give better estimates of the output sequence. The procedures for linear systems with continuous time are analogous to the above. The references by Davies and Doebelin give detailed treatment of these methods. This method provides a controlled, experimental method of identifying linear systems. It should be noted that if $\{e_i\}$ is an uncontrolled input then we can still make use of the technique by estimating both $\gamma_{ve}(k$ and $\gamma_e(s)$, though the mathematics of then finding $\{c_i\}$ is made substantially harder by not having a simple form for $\gamma_e(s)$.

12.8 Identifiability

In our study of identification we have assumed that given a set of data and a potential model we can always sensibly fit the model. This is not the case. The nature of the data and the structure of the model may prevent such a fitting process, in which case we would say that the model was not *identifiable* with the given data and information. This is a major problem with large complex models, since it is not always obvious whether a model is identifiable or not. A detailed study of identification is beyond the scope of this book. Most books on econometrics include such studies, for example the references by Koutsoyiannis (references for Part I) and by Harvey (references for Part II). The aim of this section is simply to highlight the problem by means of some examples.

Example 1

In subsection 4.8.3 we referred to identifying a regression on the basis of data taken only at three values of the regressor variable x. It is clear that though this is optimum for distinguishing between a linear and quadratic model, it is totally incapable of distinguishing between a quadratic and cubic model as both curves could pass through the same three points.

Example 2

Suppose we have an autoregressive model for a set of sales data, x_t. Thus

$$x_t = \beta x_{t-1} + e_t$$

Suppose, further, the errors are also AR (1)

$$e_t = \alpha e_{t-1} + u_t$$

where u_t are independent, zero mean. A little algebra will show that

$$x_t = (\alpha + \beta)x_{t-1} - \alpha\beta x_{t-2} + u_t,$$

so

$$x_t = \theta_1 x_{t-1} + \theta_2 x_{t-2} + u_t,$$

where

$$\theta_1 = \alpha + \beta \quad \text{and} \quad \theta_2 = \alpha\beta.$$

The latter model is a standard model, AR(2) and we can use the data to find values for θ_1 and θ_2. However, knowing θ_1 and θ_2 does not uniquely lead to values for α and β since they could be interchanged without altering θ_1 and θ_2. Thus $\theta_1 = 0.90$, $\theta_2 = 0.08$ can lead to $\alpha = 0.1$, $\beta = 0.8$ or $\alpha = 0.8$, $\beta = 0.1$. Thus the original model cannot be identified. The only way out of the dilemma is to introduce into the problem some further information that would guide us to which of the two solutions is the sensible one.

Example 3

Consider again the example of structural form and reduced form models in subsection 7.2.4. If we centre the observations about their average values to remove the constant terms, the structural form becomes

$$c_t = \alpha_1 y_t + e_{1t},$$
$$i_t = \beta_1 y_t + \beta_2 y_{t-1} + e_{2t},$$
$$y_t = c_t + i_t + g_t.$$

The reduced form becomes

$$c_t = \pi_{11} g_t + \pi_{12} y_{t-1} + f_{1t}$$
$$i_t = \pi_{21} g_t + \pi_{22} y_{t-1} + f_{2t}$$
$$y_t = \pi_{31} g_y + \pi_{32} y_{t-1} + f_{3t}$$

There are thus six parameters in the deterministic parts of the reduced form but only three in the structural form. The reduced form is said to be over-parameterized. Obviously there are relations between the two sets of parameters, which are obtained as the reduced form is derived from the structural form. Thus

$$\pi_{11} = \frac{\alpha_1 \beta_1}{1 - \alpha_1 - \beta_1}, \qquad \pi_{12} = \frac{\alpha_1}{1 - \alpha_1 - \beta_1},$$

$$\pi_{21} = \frac{\beta_2(1 - \alpha_1)}{1 - \alpha_1 - \beta_1}, \qquad \pi_{22} = \frac{\beta_1}{1 - \alpha_1 - \beta_1},$$

$$\pi_{31} = \frac{\beta_2}{1 - \alpha_1 - \beta_1}, \qquad \pi_{32} = \frac{1}{1 - \alpha_1 - \beta_1}.$$

Notice that $\beta_1 = \pi_{11}/\pi_{12}$ and also $\beta_1 = \pi_{22}/\pi_{32}$. If we knew the values of π_{ij} exactly this mathematical result would be just a statement of fact. However, in practice we may only have estimates of the values of π_{ij} so we will have two alternative and almost certainly inconsistent ratios each estimating β_1. Further it is evident that if we used as formula $\alpha_1 = \pi_{12}/\pi_{32}$, $\beta_1 = \pi_{11}/\pi_{12}$, $\beta_2 = \pi_{31}/\pi_{32}$, then a change of π_{11}, π_{12}, π_{31}, π_{32} by any multiplier will leave α_1, β_1 and β_2 unaltered. So an infinity of reduced form equations are consistent with the same structural equations.

Example 4
Consider again the glucose example of section 9.2, where g and h related to the amount of glucose and hormones in the blood at time t. The natural form of the model was

$$\frac{dg}{dt} = -\alpha g - \beta h,$$

$$\frac{dh}{dt} = -\gamma h + \delta g.$$

These equations gave rise to an equation

$$\frac{d^2g}{dt^2} + (\alpha + \gamma)\frac{dg}{dt} + (\alpha\gamma + \beta\delta)g = 0.$$

The parameters in this equation can be reexpressed to give a more convenient equation

$$\frac{d^2g}{dt^2} + 2\xi\omega\frac{dg}{dt} + \omega^2 g = 0,$$

which has as a usual solution assuming $\xi < 1$

$$g = e^{-\xi\omega t}(A\cos\omega_0 t + B\sin\omega_0 t), \text{ where } \omega_0 = \omega(1 - \xi^2)^{\frac{1}{2}}$$

The parameters A and B in this solution form arise in the solution of the equation and relate to the initial values of g and its rate of change. The other parameters satisfy the relations

$$\omega^2 = \alpha\gamma + \beta\delta$$

$$\xi = (\alpha + \gamma)/2\omega$$

Thus there are two parameters in the solution form and four related parameters in the natural form. So even if we know ω and ξ exactly we cannot find α, β, γ or δ. To find these we would have to have direct information on the elements in the equations of the natural form model.

Example 5
Examples 3 and 4 considered identification problems relating one form of a model to another. There may be identification problems within structural equations themselves. Suppose the structural form of Example 3 was slightly modified by allowing c_t to depend on y_{t-1} as well as y_t and setting the government input, g_t, at zero. The first equation in Example 3 then becomes

$$c_t = \alpha_1 y_t + \alpha_2 y_{t-1} + e_{1t}.$$

Substituting $i_t = y_t - c_t$ from the third equation into the second gives:

$$y_t - c_t = \beta_1 y_t - \beta_2 y_{t-1} + e_{2t},$$

which can be rewritten in the form

$$c_t = \gamma_1 y_t - \beta_2 y_{t-1} - e_{2t}$$

This is identical in general form with the first equation. There is no reason however why α_2 should equal $-\beta_2$, yet given any set of data we must obtain the same value for the coefficient of y_{t-1} in these two equivalent equations. Thus from the viewpoint of the data there can be only one relation between c_t and (y_t, y_{t-1}), but yet the structural equations imply two different relations. There exist standard rules, given in the two previous references, that enable each of the structural equations to be checked to see whether it can be identified.

Example 6

Consider a two-variable regression

$$y_i = \beta_0 + \beta_1 x_{1i} + \beta_2 x_{2i} + e_i$$

Suppose that there is present an additional relation between the regressor variables:

$$x_{1i} = \alpha x_{2i}$$

The model is thus equivalent to

$$y_i = \beta_0 + (\beta_1 \alpha + \beta_2) x_{2i} + e_i$$
$$= \beta_0 + \beta^* x_{2i} + e_i.$$

Thus, even if we had error free data, $e_i = 0$, we could find β_0, β^* and α but we could not find β_1 and β_2 from these. The occurrence of a high, but not perfect, linear correlation between regressor variables is by no means uncommon. The consequence of this correlation is that though values for β_1 and β_2 could be found they would be highly unreliable. This phenomena is called *Multicollinearity*. Ways out of the problem are to use only x_2, or to create a new variable combining x_1 and x_2.

Part IV

Estimation

Chapter 13

Empirical model fitting

13.1 Approaches to fitting models to data

The objective of this Part of the book is to consider how an assumed model may be fitted to a set of data.

Example 1
In Chapter 1 we considered modelling a set of data (Table 1.1) by a simple linear model

$$y_i = \theta x_i + e_i$$

As we saw in that example a reasonable fit between model and data was achieved by the numerical value 1.3 being assigned to θ. This value, written as $\hat{\theta} = 1.3$, we called an estimate of θ. The quantity $\hat{\theta}$ expressed as a rule for calculating an estimate, such as

$$\hat{\theta} = \frac{\sum xy}{\sum x^2},$$

is called an estimator. With a sensible estimator the fitted line will pass closely to the individual points, so we shall want the residuals, $\hat{e}_i$, to be small. The residuals are measures of difference between fitted model and data, so one approach to fitting the model would be to minimize the overall difference as measured, for example, by $\Sigma \hat{e}_i^2$. For our model

$$\hat{e}_i = y_i - \hat{\theta} x_i$$

where $\hat{\theta}$ is an estimator of θ, and hence

$$\Sigma \hat{e}_i^2 = \Sigma (y_i - \hat{\theta} x_i)^2$$

This can be written as

$$\Sigma \hat{e}_i^2 = \Sigma y^2 - \frac{(\sum xy)^2}{\sum x^2} + \left[\sum xy - \hat{\theta} \sum x^2 \right]^2 / \sum x^2$$

which is minimum when the term in [　] is set at zero

i.e. for
$$\hat{\theta} = \sum xy / \sum x^2.$$

which gives $\hat{\theta} = 1.289$ for the data, see Table 13.1. So the model is *Fitted* by defining a criterion and then choosing an estimator for the parameter that satisfies this criterion.

The problem of fitting presents itself in a wide variety of forms and can be approached with a wide variety of methods. At this stage it is useful to seek a classification of the methods. The classification to be used is not

Table 13.1 Residuals for road distance model

(a)	Linear distance x	Road distance y	Fittd road distance $\hat{y}$	Residual $\hat{e}$
	4.8	6.5	6.2	0.3
	5.0	6.5	6.4	0.1
	8.3	9.5	10.7	−1.2
	9.5	10.7	12.2	−1.5
	9.8	11.7	12.6	−0.9
	11.4	18.4	14.7	3.7
	11.8	19.7	15.2	4.5
	12.1	14.2	15.6	−1.4
	12.1	16.6	15.6	1.0
	14.6	16.3	18.8	−2.5
	15.2	17.2	19.6	−2.4
	18.0	26.5	23.2	3.3
	19.0	25.6	24.5	1.1
	21.6	28.8	27.8	1.0
	21.7	25.7	28.0	−2.3
	22.0	29.0	28.4	0.6
	23.0	29.4	29.6	−0.2
	26.5	31.2	34.2	−3.0
	28.0	23.1	36.1	−3.0
	28.2	40.5	36.4	4.1

(b)　Calculations for future reference

$$\sum x = 322.6 \qquad \sum y = 417.1$$
$$\sum x_i^2 = 6226.38 \qquad \sum y^2 = 10453.71$$
$$\sum x_i y_i = 8026.25 \qquad \hat{\theta} = \frac{\sum xy}{\sum x^2} = 1.289, \ \hat{y}_i = \hat{\theta} x_i$$
$$\sum \hat{e}^2 = 107.11$$
$$\hat{\sigma}_e^2 = \sum \hat{e}^2 / (n-1) = (2.37)^2$$

(Standard error of $\hat{\theta}$)$^2 = \dfrac{\hat{\sigma}_e^2}{\sum x_i^2} = (0.030)^2$

intended as perfect or all embracing, but rather one that covers the most common methods and is helpful in describing their nature and interrelations. The distinction made between Empirical and Conceptual approaches provides an initial classification. In an empirical approach to fitting, we examine our model and our data and consider how we can make the two fit closely together. To do this we need criteria that measure this closeness, *Criteria of Fit*. There are three major groups of such criteria

(a) The first group is based on establishing equality between measures of the shape of the model and the analogous measures of the shape of the data. For example the mean of the fitted distribution could be set equal to the mean of the data.
(b) The second group is based on using measures of the discrepancy between the fitted model and the data. In Example 1 the residual sum of squares was used as the measure of discrepancy.
(c) The final group of criteria is based on the idea that the model in some sense explains the structure of the data. If we can devise measures of how well the data is explained by the model, then we can fit the model in such a way as to maximise such a measure. Likelihood provides such a measure of explanatory ability.

Note that, though in our empirical approaches we started by considering fitting the complete model, we moved naturally to considering estimators of the individual parameters. A conceptual approach to fitting similarly concentrates on the estimators and asks, in conceptual rather than empirical terms, 'What constitutes a good estimator?', 'What properties should a good estimator have?'. Answering these questions leads to the consideration of the expectations and variances of estimators. We must also examine whether the ultimate applications of the model can contribute to our choice of methods of fitting. These approaches are considered in the next chapter.

Finally if we take a more eclectic approach we can consider (in Chapter 15) ways of bringing the empirically and conceptually oriented methods together.

13.2 Methods of shape equalization

The methods under this heading are concerned particularly with distributional models. The parameters in distributional models clearly influence the shape of the distribution and indeed they often have a direct interpretation in terms of the shape. For example in the Normal distribution, $N(\mu, \sigma^2)$, μ is the middle point and σ the distance from the middle to the points of inflexion on the sides (Figure 3.3). Given a set of data we may devise sample statistics to measure analogous aspects of the shape of the distribution of the sample. For example, we may use the sample mean, median, or mode as measures of the position of the data on its axis. Wherever we make such an analogy between an aspect of the shape of the distribution and the shape implied by the data we can choose the parameter values to equalize these two quantities. Suppose that the distribution has parameters, α and β, and $S_1(\alpha, \beta)$, $S_2(\alpha, \beta)$ are two measures of the shape of the distribution. Then if s_1 and s_2 are the numerical values of the analogous measures of the shape of the data we can find estimates $\hat{\alpha}$ and $\hat{\beta}$ by solving

$$S_1(\hat{\alpha}, \hat{\beta}) = s_1$$
$$S_2(\hat{\alpha}, \hat{\beta}) = s_2$$

In illustrating this approach we consider two sets of examples, those in which shape is measured purely in terms of moments and those where other measures are used.

13.2.1 The method of moments

The shape of both distribution and data can be defined by their respective means, variances, skewness, etc. in other words by their moments (see sections 10.2 to 10.4 for sample moments and section 2.5 for population moments). If there is only one parameter its estimator can be chosen to make the population mean equal to the sample mean. In general we choose the simplest moments to equate.

Example 1
For the exponential distribution

$$f(x) = \lambda e^{-\lambda x} \qquad x \geqslant 0, \lambda > 0$$

the population mean, the first moment, is $E(X) = 1/\lambda$. The sample mean is $\bar{x}$, so the estimator, $\hat{\lambda}$, of λ that makes these two the same is

$$\hat{\lambda} = 1/\bar{x} = n \left/ \sum_1^n x_i \right.$$

The fitted model is thus

$$\hat{f}(x) = \frac{1}{\bar{x}} e^{-x/\bar{x}} \qquad x \geqslant 0$$

If there are two parameters we can set them so that both mean and variances are the same for the fitted distribution as for the sample.

Example 2
In the Systems Dynamic models, discussed in section 7.6, use is often made of models for time delays based on the exponential distribution

$$f(t) = \lambda e^{-\lambda t}.$$

A first order delay is modelled by this distribution. A Kth order time delay, T, is regarded as made up of the sum of K independent exponential delays, t_i,

$$T = \sum_1^K t_i.$$

This quantity has a distribution called an Erlang distribution. The expectation and variance of the exponential are $1/\lambda$ and $1/\lambda^2$ respectively. Hence from the laws of addition the expectation and variance of T are K/λ and K/λ^2 respectively. Given several observations on delays $T_1, T_2, \ldots, T_n$, on a system with unknown λ and K we can calculate the average delay time $\overline{T}$ and the standard deviation s_T. The method of moments thus gives

1st moment $\qquad \dfrac{\hat{K}}{\hat{\lambda}} = \overline{T}$

2nd moment about the mean $\dfrac{\hat{K}}{\hat{\lambda}^2} = s_T^2.$

Hence

$$\hat{K} = \overline{T}^2/s_T^2$$

As K is an integer we suitably round $\hat{K}$ to an integer to estimate the order of the delay.

Example 3

For a Normal distribution, $N(\mu, \sigma^2)$, the first and second moments about the mean of the sample are

$$m_1 = \bar{x} = \sum_1^n x_i/n,$$

$$m_2 = s'^2 = \sum_1^n (x_i - \bar{x})^2/n,$$

For the population the corresponding moments are

$$\mu_1 = \mu, \qquad \mu_2 = \sigma^2$$

so the method of moments estimators are

$$\hat{\mu} = \bar{x}, \qquad \hat{\sigma}^2 = s'^2.$$

In general then the method of moments provides a series of estimating equations by using the analogous nature of the population moments $E(X^k)$, $k = 1, 2, \ldots$, and the sample moments, $(1/n)\Sigma x^k$. The parameters appear in the population moments $E(X^k|\theta)$ so the estimating equations can be written as

$$E(X^k|\hat{\theta}) = \frac{1}{n}\Sigma x^k, \ k = 1, 2, \ldots .$$

176

The number of equations used equals the number of unknown parameters. The moments used may be those about the origin or about the mean, whichever is most convenient.

The major problem with the method of moments is that higher moments have very large and increasing variability and hence unreliability as estimates of the population higher moments. Estimating three or more parameters requires use of $m_3, \ldots$, and this gives rise to unreliable estimates.

The method can be applied using covariances and correlations as well as means and variances.

Example 4
In the discussion of the autoregressive model the Yule–Walker equations were obtained that related the parameters ϕ to the population autocorrelation coefficients, ρ_k. If these population values are replaced by their estimates, r_k, obtained from the data then we have a set of simultaneous equations for the unknown ϕ which can be solved to give method of moments estimators $\hat{\phi}$.

We conclude this section by applying the method to a somewhat more complex modelling problem.

Example 5
The position of a pulse appearing on a radar screen should correspond, on a suitable scale, to the position of the point where the radar pulse originated, see Figure 13.1. A number of sources of error prevent an exact correspondence. Suppose we are estimating the length, l, of a passing ship from the pulse data giving coordinates y_i as estimates of source points. As a model we regard each y_i as being the true coordinate x_i on the ship together with an error e_i. For simplicity we assume that e_i has zero mean and normal distribution and that its variance, σ^2, is known, being mainly a property of the equipment and the known range of the ship. The x_i will be taken to be uniformly distributed over the length of the ship since they are the reflecting points for a uniform beam of radar pulses. Thus we have

$$y_i = x_i + e_i, \qquad x_i \cap U(k, k+l).$$
$$e_i \cap N(0, \sigma^2).$$

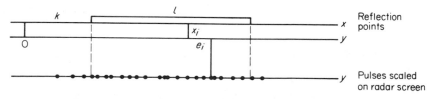

Figure 13.1 Pulsed radar reflections from a ship

where k is the coordinate of, say, the stern of the ship relative to some purely arbitrary origin. It should be noted that we cannot use the stern of the ship as origin, as the occurrence of negative errors, with values of x_i near the stern, means that we do not know on the radar screen exactly where the stern is. If we take expectations we have

$$E(Y_i) = E(X_i) + E(E_i) = k + l/2$$

Thus the average $\bar{y}$ estimates the position of the middle of the ship but does not tell us its length. The assumption that x_i and e_i are independent values leads to

$$V(Y) = V(X) + V(E)$$

$$= \frac{l^2}{12} + \sigma^2$$

using the variance of the uniform distribution of $l^2/12$. The sample estimate of $V(Y)$ is s_y^2, which can be measured automatically from the incoming data. Thus the method of moments estimator of l is obtained by equating $V(Y)$ and s_y^2 which leads to

$$\hat{l} = 12(s_y^2 - \sigma^2)^{\frac{1}{2}}.$$

13.2.2 Other methods based on population-sample analogues

The method of moments is based essentially on the analogous nature of population moments $E(X^k)$ and sample moments $(1/n)\Sigma x_i^k$. There may be other properties of a distribution for which an analogy may give a simple estimator. These will be illustrated by a number of examples.

Example 1
Consider again the uniform distribution $U(0, \theta)$

$$f(x) = \begin{matrix} \dfrac{1}{\theta} & 0 \leqslant x \leqslant \theta \\ \\ 0 & \text{otherwise} \end{matrix}$$

Pure intuition suggests that the largest observation, $x_{(n)}$, contains the best information about the top end of the distribution, θ. Now it can be shown that the expected value of $x_{(n)}$ is $n\theta/(n+1)$. If we equate the expected value of $x_{(n)}$ in the population with the observed value in the sample we obtain an estimation equation

$$x_{(n)} = n\hat{\theta}/(n+1)$$

leading to

$$\hat{\theta} = \left(1 + \frac{1}{n}\right) x_{(n)}.$$

Example 2

In section 11.2 we mentioned the use of normal probability paper. On this paper the population is represented by a straight line, which has on the x-axis the value μ at the 50% point on the probability axis and the values $\mu - \sigma$ and $\mu + \sigma$ at, approximately, the 16% and 84% points (see Figure 11.6). A sample will give a set of points which, assuming our model is reasonable, will be approximately a straight line. If by eye, or by some formal method, we draw the sample line we may use the population-sample analogy to read off from the 50% point and the 16%–84% difference estimates of μ and σ, giving $\hat{\mu} = 10$ and $\hat{\sigma} = 0.9$ approximately for Figure 11.6. Other probability papers can be used to provide estimators for other distributions.

Example 3

A distribution of wide use but considerable complexity is the Weibull distribution with p.d.f.:

$$f(x) = (\gamma x^{\gamma-1}/\beta^{\gamma}) \exp -(x/\beta)^{\gamma}, \quad 0 \leqslant x \leqslant \infty, \ \beta > 0, \ \gamma > 0.$$

Most standard methods of estimation lead to equations requiring approximate solution. However, the pth percentile, x_p, is given by the relatively simple equation

$$p = 1 - \exp\{-(x_p/\beta)^{\gamma}\}.$$

Thus if two percentiles, say the 10% and 90%, are obtained from a cumulative plot, or from the ordered data, then two equations are obtained

$$0.1 = 1 - \exp\{-(x_{0.1}/\hat{\beta})^{\hat{\gamma}}\}$$
$$0.9 = 1 - \exp\{-(x_{0.9}/\hat{\beta})^{\hat{\gamma}}\}$$

These equations are based on the analogy of the population and sample percentiles. A little arithmetic will show that they can be simply solved to give $\hat{\beta}$ and $\hat{\gamma}$. A re-examination of the previous example shows that the method applied there was just a graphical version of this method.

Example 4

It may happen that a simple feature of the shape corresponds to a complicated function of the parameters. For example consider a distribution to be considered in more detail in Chapter 19. This distribution is a Poisson Distribution with no zero observation allowed. The distribution is given by

$$Pr(R = r) = e^{-\mu}\frac{\mu^r}{r!} \bigg/ (1 - e^{-\mu}) \qquad r = 1, 2, \ldots.$$

The expectation of R is

$$E(R) = \frac{\mu}{1 - e^{-\mu}},$$

so the method of moments gives an equation that has no explicit solution.

Suppose, however, we look for some other function of the shape that relates more simply to the parameter. Such a function for the above population is

$$\frac{2\,Prob(R = 2)}{Prob\,(R = 1)} = \mu.$$

For the data, Table 19.2, the corresponding sample quantity is

$$\frac{2 \times (\text{frequency of } R = 2)}{(\text{frequency of } R = 1)} = \frac{2 \times 133}{436} = 0.610,$$

so equating the analogous quantities gives directly

$$\hat{\mu} = 0.610.$$

Clearly this method makes no use of $Prob\,(R = 3)$, etc., or of the frequencies for $R = 3, 4$, etc. However, from the data it can be seen that $R = 1$ and 2 are the dominant terms and we would thus expect to get a reasonable estimator from this approach. This example raises the two ideas (a) we can devise methods for obtaining explicit estimators suitable for specific distribution functions (b) the quality of estimates may depend on the specific data.

Example 5
The following data give the weight in grams of a small animal over the first twelve days of growth measured at 12.00 each day

25.9, 27.1, 32.6, 37.8, 38.5, 46.0, 49.1, 57.3, 62.2, 71.0, 82.6, 95.5

The suggested model is

$$x_t = \alpha e^{\beta t} + e_t.$$

If we divide the data into two equal sections

$$x_1\ x_2, \ldots, x_n \mid x_{n+1}, x_{n+2}, \ldots, x_{n+n},$$

then the sums of the two sections are, from the model:

$$S_1 = \alpha(e^\beta + e^{2\beta} + \ldots + e^{n\beta}) + \sum_1^n e_i$$

$$S_2 = \alpha e^{n\beta}(e^\beta + e^{2\beta} + \ldots + e^{n\beta}) + \sum_{n+1}^{2n} e_i.$$

Comparing expectations we have

$$E(S_2) = e^{n\beta} E(S_1).$$

Replacing the expectations by the corresponding sample sums gives the relation

$$\sum_{n+1}^{2n} x_i = e^{n\beta} \sum_1^n x_i$$

which we can solve for $\hat{\beta}$. For the data we have $417.7 = e^{6\beta} \times 207.9$ taking logs and solving gives

$$\hat{\beta} = 0.116.$$

The value of α can now be estimated using

$$E(S_1 + S_2) = \alpha(e^\beta + e^{2\beta} + \qquad + e^{12\beta})$$

substituting the observed $S_1 + S_2$ on the left-hand side and the estimated $\hat{\beta}$ on the right gives $\hat{\alpha} = 22.62$.

13.3 Methods of minimum deviation

13.3.1 General methods

In the introduction to this chapter we illustrated the use of criteria based on measures of the closeness of the data points to the fitted model. Such methods can be called *Methods of Minimum Deviation*. There are many such methods that vary according to the type of model being fitted and the way in which deviation is assessed. As a simple introduction consider the following.

Example 1
The following data represents a series of measurements made on a concentration of a chemical in solution (in grams/litre).

10.4, 9.3, 11.2, 10.1, 10.8, 9.7, 9.7, 8.2, 10.2, 10.6, 9.5, 10.0, 10.1, 10.2, 10.3, 9.2, 10.4, 8.8, 11.0, 8.7

Denoting the data by x_i, $i = 1, 2, \ldots, n$ we use a model

$$x_i = \mu + e_i,$$

where μ is a parameter indicating the 'centre' of the data and e_i are independent random variables of zero mean and constant variance. To fit this model an estimate $\hat{\mu}$ of μ must be found. The deviation of the data from the fitted μ is $\hat{e}_i = x_i - \hat{\mu}$. A close fit will occur if the quantities $\hat{e}_1 \hat{e}_2, \ldots, \hat{e}_n$ are all small. It is sensible to produce a single measure of the overall deviation. The tradition in science, going back to the eighteenth century, is to use as a measure the sum of squares of the residuals

$$S = \sum_{i=1}^{n} \hat{e}_i^2.$$

The use of $\hat{e}^2$ has the merit that the sign of $\hat{e}_i$ becomes irrelevant and that, because of the squaring, $\hat{\mu}$ is chosen to particularly avoid large $\hat{e}_i$ which will make excessively large contributions to S. Having chosen S as the measure of deviation we now choose $\hat{\mu}$ to minimize this criterion of fit. The method is thus called the *Method of Least Squares*. For the example we thus have

$$S = \sum_{i=1}^{n} (x_i - \hat{\mu})^2.$$

If we denote the average by $\bar{x}$ it will be seen that S can be re-expressed as

$$S = \sum_{i=1}^{n} [(x_i - \bar{x}) + (\bar{x} - \hat{\mu})]^2$$

$$= \sum (x_i - \bar{x})^2 + n(\bar{x} - \hat{\mu})^2,$$

since

$$\sum (x_i - \bar{x}) = 0.$$

From this it is clear that S can be minimized by setting $\hat{\mu} = \bar{x}$. Thus the Least Squares estimate of μ is $\bar{x} = 9.92$ grams/litre.

Example 2

Suppose we now change our area of application and assume that the data of Example 1 gives the price per kilo of a household product at a number of different shops. If we followed the above argument and took $\hat{\mu} = 9.9$, rounding to the nearest unit then the residuals for the first and last shops would be 0.5 and -1.2. The use of squared residuals would take the contribution of these deviations as 0.25 and 1.44. Thus just over twice the price difference is counted as over five times as important. This is clearly not very appropriate. The natural measures of deviation in this case is just the magnitudes, $|\hat{e}_i|$, of the residuals and the overall measure is

$$S_a = \sum_{i=1}^{n} |\hat{e}_i|,$$

which for our data is

$$S_a = \sum_{i=1}^{n} |x_i - \hat{\mu}|.$$

The value of $\hat{\mu}$ that minimizes this is the median, the middle value of the data, in this case, $\hat{\mu} = 10.1$. This can be seen by pairing the ordered observations $(x_{(1)}, x_{(n)})$, $(x_{(2)}, x_{(n-1)})$, etc.

Note that

$$|x_{(1)} - \hat{\mu}| + |x_{(n)} - \hat{\mu}| = (\hat{\mu} - x_{(1)}) + (x_{(n)} - \hat{\mu}) = x_{(n)} - x_{(1)},$$

which does not depend on $\hat{\mu}$, provided $\hat{\mu}$ lies in $(x_{(1)} - x_{(n)})$. The same argument applies for each pair until the median is reached. It is then apparent that if $\hat{\mu}$ moves from the median position it must increase S_a.

Example 3

Let us extend these ideas to the simple linear model of Chapter 1, namely

$$y_i = \theta x_i + e_i.$$

If we have an estimator $\hat{\theta}$ of θ then the fitted model will be

$$\hat{y} = \hat{\theta}x,$$

where $\hat{y}$ denotes the fitted, estimated, value of y. The residuals, $\hat{e}_i$ will be

$$\hat{e}_i = y_i - \hat{\theta}x_i.$$

The two measures, criteria, of closeness considered in the previous examples were

$$C_1 = \sum |\hat{e}_i|$$
$$C_2 = \sum \hat{e}_i^2.$$

The criterion C_2 is the easiest to use and estimates obtained from it are the Least Squares estimates. For this the criterion becomes

$$C_2\!: \sum_{n=1}^{n} \hat{e}_1^2 = \sum_{i=1}^{n} (y_i - \hat{\theta}x_i)^2.$$

To find the $\hat{\theta}$ that minimizes this, i.e. that gives the closest fitting line as defined by C_2, we can use an algebraic method as was done at the beginning of this chapter. The more usual method requires the use of elementary calculus. Differentiating with respect to $\hat{\theta}$ and equating to zero gives:

$$\frac{d}{d\hat{\theta}} \sum (y_i - \hat{\theta}x_i)^2 = -2 \sum x_i (y_i - \hat{\theta}x_i) = 0 \text{ for a minimum value,}$$

so

$$\hat{\theta} = \sum x_i y_i / \sum x_i^2.$$

If we specify a value for $\hat{\theta}$ the criterion C_1 is

$$C_1 = \sum |y_i - \hat{\theta}x_i|.$$

A simple computer program can evaluate this in a fraction of a second for even very large n, so that it is simple to try a range of values of $\hat{\theta}$ and evaluate C_1 at each $\hat{\theta}$. The program of Table 13.2 shows such a program in the language BASIC. This is not the most elegant or efficient program. It was written in about 20 minutes by an inexperienced programmer to make the point that error criteria are very simple to evaluate when there are only one or two parameters. Suppose we decide that errors get very expensive very quickly then we might use a symmetric cubed error $\sum |\hat{e}^3|$ criteria. In which case line 60 of the program, becomes

$$60 \quad A = A + E \uparrow 3,$$

$E \uparrow 3$ denoting E^3. Table 13.3 shows the values of θ and of C_1 calculated by the program for a series of values seeking the optimum $\hat{\theta}$. About a minute's search yielded $\hat{\theta} = 1.278$ as the optimum value to four significant figures. It will be noted that this estimator is quite close, both numerically and in terms of the value of $\sum |\hat{e}|$, to the least squares estimator. This is a

Table 13.2 A rough program for the interactive optimization of $\sum|\hat{e}|$, A in the program, B being the positive parameter estimate $(\hat{\theta})$

5	DIM X(500), Y(500)
10	INPUT N
20	FOR I = 1 TO N
30	INPUT X(I), Y(I)
35	NEXT I
37	INPUT B
38	A = 0
39	FOR I = 1 TO N
40	E = Y(I) − B*X(I)
50	E = ABS(E)
60	A = A + E
70	NEXT I
80	PRINT B, A
85	INPUT B
90	IF B < 0 THEN 100
95	GO TO 38
100	END

Comment: The program requires first N, the number of observations ($\leqslant 500$), then the data $(X(1),\ Y(1)),\ \ldots,\ (X(N),\ Y(N))$, then a series of values of the slope, B, for each of which it prints B and the value of $\sum|\hat{e}|$. The program stops if a negative B is input.

Table 13.3 Exploratory estimation

Value of $\hat{\theta}$ tried	Value of $\sum|\hat{e}|$ given by program
1.0	94.5
2.0	228.1
1.5	73.4
1.3	38.38
1.2	40.46
1.4	50.42
1.25	38.90
1.29	38.19
1.28	38.044
1.27	38.276
1.282	38.0736
1.279	38.0292
1.278	38.0264
1.2785	38.0218

consequence of the reasonably symmetric distribution of the $\hat{e}_i$ values and the absence of particularly extreme values, both of which effect $\Sigma \hat{e}^2$ more than $\Sigma |\hat{e}|$.

We have examined two criteria based on the residuals. There is an infinite number of possibilities. In Table 13.4 six criteria of the form $\Sigma |\hat{e}_i|^k$ are evaluated for various values of $\hat{\theta}$. The data used had most values corresponding to θ of about 1.5 but had one value corresponding to $\theta = 2.0$. It can be seen that the criterion is minimum for larger $\hat{\theta}$ as k increases. This is because the large $\hat{e}$, associated with the one discrepant observation, has a greater influence when the larger power of $|\hat{e}|$ is used in the criterion. This suggests that the use of large k is not wise unless the situation is such that a large $\hat{e}$ does have a disproportionately bad effect.

13.3.2 The method of least squares

Historically the first criterion used, by astronomers in the eighteenth century, was $\Sigma |\hat{e}|$. However, it was soon found that, though this made a great deal of intuitive sense in the problems being examined, it was very difficult to do the calculations; whereas $\Sigma \hat{e}_i^2$ was much more amenable to calculation. So the method of least squares was born and ever since has been used almost as though it were the only possible approach. With the advent of the computer we can now evaluate different criteria with almost equal ease. So we now have the practical possibility of evaluating and using the criteria most suitable for any particular practical problem. It is apparent from the literature that, with over 200 years of study and use behind it, the least squares criterion still dominates the theory and the computer software of model fitting. As we continue our study the value and elegance of Least Squares should become more apparent; however, that

Table 13.4 Values of $\sum \hat{U}_i$ for various values of $\hat{\theta}$
in the model $y = \theta x + e$, where $\hat{U}_i = |\hat{e}_i|^k$

			Form of $\hat{U}_i$			
$\hat{\theta}$	$\sqrt{e_i}$	$|e_i|$	$|e_i|^{1.5}$	e_i^2	$|e_i|^3$	e_i^4
1.500	3.04	3.20	4.62	7.36	19.69	53.15
1.525	3.00	3.18	4.43	6.79	17.12	43.98
1.55	3.31	3.35	4.36	6.33	14.85	36.09
1.60	4.19	3.90	4.46	5.75	11.25	23.87
1.65	4.73	4.45	4.79	5.63	9.05	16.08
1.70	5.16	5.00	5.30	5.96	8.38	12.68
1.75	5.51	5.55	5.98	6.75	9.38	13.98
1.80	5.81	6.10	6.81	7.99	12.21	20.61

must never be an excuse for using it in a situation where $\Sigma \hat{e}_i^2$ is not a suitable measure of the effects of error in the fitted model.

We now consider some further examples of using Least Squares. The first is the most common of all, namely the fitting of a straight line to a set of data.

Example 1

The basic linear model is

$$y_i = \beta_0 + \beta_1 x_i + e_i.$$

Thus the criterion is

$$\Sigma \hat{e}^2 = \Sigma (y_i - \hat{\beta}_0 - \hat{\beta}_1 x_i)^2$$

To find the minimum, this is differentiated with respect to each parameter and equated to zero. Differentiating with respect to β_0 and equating to zero gives

$$-2 \Sigma (y_i - \hat{\beta}_0 - \hat{\beta}_1 x_i) = 0.$$

This can be written as

$$\Sigma y_i = n \hat{\beta}_0 + \hat{\beta}_1 \Sigma x_i.$$

Differentiating with respect to $\hat{\beta}_1$ and equating to zero gives

$$-2 \Sigma x_i (y_i - \hat{\beta}_0 - \hat{\beta}_1 x_i) = 0.$$

This can be written as

$$\Sigma x_i y_i = \hat{\beta}_0 \Sigma x_i + \hat{\beta}_1 \Sigma x_i^2.$$

The two basic equations obtained above are called the *Normal Equations* and they give two simultaneous equations that can be solved for $\hat{\beta}_0$ and $\hat{\beta}_1$. Note that the first normal equation can be written as

$$\bar{y} = \hat{\beta}_0 + \hat{\beta}_1 \bar{x},$$

so the fitted line goes through the average point $(\bar{x}, \bar{y})$. Notice also that the two normal equations can be expressed as

$$\Sigma \hat{e}_i = 0 \text{ and } \Sigma x_i \hat{e}_i = 0.$$

The implication and interpretation of those two expressions is, that the average residual is zero and that the correlation between residual and regressor variable is zero. Both of these are intuitively reasonable requirements of residuals, though note that the first does not apply if the regression model is through the origin, i.e. if $\beta_0 = 0$.

If we had made, centred, our measurements (x_i', y_i') from $(\bar{x}, \bar{y})$, i.e. $y_i' = y_i - \bar{y}$, the first normal equation would give $\hat{\beta}_0' = 0$ and the second would give

$$\hat{\beta}_1' = \Sigma (x_i' y_i') / \Sigma (x')^2$$

The fitted line is then

$$y' = \hat{\beta}_1' x'.$$

This is

$$y - \bar{y} = \hat{\beta}_1 (x - \bar{x}), \text{ since } \hat{\beta}'_1 = \hat{\beta}_1.$$

Example 2
The generalization of the simple linear model is the *General Linear Model*

$$y_i = \beta_0 + \beta_1 x_{i1} + \beta_2 x_{i2} + \ldots + \beta_p x_{ip} + e_i.$$

The estimation of $\beta_0, \beta_1, \ldots, \beta_p$ by the method of Least Squares is exactly parallel to the simple case and the Normal Equations can be analogosly expressed as

where
$$\sum \hat{e}_i = 0, \quad \sum \hat{e}_i x_{i1} = 0, \quad \sum \hat{e}_i x_{i2} = 0, \ldots, \sum \hat{e}_i x_{ip} = 0,$$

$$\hat{e}_i = y_i - \hat{\beta}_0 - \hat{\beta}_1 x_{i1} - \ldots - \hat{\beta}_p x_{ip}.$$

There are thus $p + 1$ simultaneous equations for the $p + 1$ estimators.

In both the simple and general linear regression models the method of Least Squares has provided the means of estimating the regression coefficients β_j. It does not provide an estimator of the hidden parameter, the error variance, σ_e^2. An estimator can be obtained by the method of moments since the sample variation about the regression line is given by

$$s'^2 = \frac{1}{n} \sum \hat{e}^2$$

which can be regarded as the estimator of σ^2. It can be shown that for the General Linear Model

$$E(s'^2) = \frac{n - p}{n} \sigma^2,$$

so a modified estimator is often used, namely

$$s^2 = \frac{1}{n - p} \sum \hat{e}^2,$$

since this has expectation σ^2.

In Sections 6.1 and 6.2 there was an examination of the assumptions normally made about the error and regressor variables in the linear regression model. We may use Least Squares to estimate the regression coefficients, β, without making any of these assumptions. If, however, these assumptions do not hold, the estimators obtained may turn out to be misleading and to have undesirable properties. We usually find these failures through techniques such as residual analysis after having fitted initially by Least Squares. In these cases we make adjustments for the failure of the assumptions at a second stage. Some of these adjustments are discussed in Part VI. Sometimes however a problem can be identified by examining the model itself. In this case we may need to modify the method of estimation.

Example 3

Consider again the simultaneous equation model discussed in Section 7.2.4 and illustrated in Figure 7.6. If the observations were centred about their means the equation for consumption becomes the simple regression through the origin:

$$c_t = \theta y_t + e_t.$$

For the purpose of illustration consider estimating θ from data on c and y. The Least Squares estimator is

$$\hat{\theta} = \frac{\sum c_t y_t}{\sum y_t^2}$$

which, using the model, can be written as

$$\hat{\theta} = \theta + \frac{\sum e_t y_t}{\sum y_t^2}.$$

If we can treat the y_t as constant, or at least uncorrelated with e_t, then the expectation of the last term is a multiple of $E(e)$ which is zero and so $\hat{\theta}$ has expectation θ. If, however, y_t and e_t are correlated we cannot make use of the rule that expectations of products are products of expectations, so the last term is not zero and will not even tend to zero for large amounts of data. An examination of Figure 7.6(c) shows that e_t influences y_t, so there is in fact a correlation. A method that avoids this problem is to use a proxy variable, z_t, for y_t, that is uncorrelated with e_t but which has a high correlation with y_t. The variable z_t is called an *Instrumental Variable*. The modified estimator of θ is

$$\tilde{\theta} = \frac{\sum c_t z_t}{\sum y_t z_t},$$

called the *Instrumental Variable Estimator*. Again substituting the model gives

$$\tilde{\theta} = \theta + \frac{\sum e_t z_t}{\sum y_t z_t}$$

Though the expectation of the final term is not necessarily zero, the lack of correlation between e and z ensures that it will tend to zero for large amounts of data. The choice of instrumental variable is very much a function of the problem and model being studied. Often a variable already in the equations may be used. For example e_t will intuitively be independent of y_{t-1} and we might expect a high correlation between y_t and y_{t-1}. Thus y_{t-1} is a candidate instrumental variable. Sometimes a new instrumental variable may be created by a suitable weighted average of a number of possible candidates. The weights are chosen to give the maximum correlation with the regressor for which the instrumental variable is required.

13.3.3 Fitting non-linear models*

In Section 6.4 we referred to models that could be transformed to linearity. When this is the case the result of applying least squares to the transformed model will be to give a set of estimates $\hat{\beta}_i$ and their standard errors, s_{β_i}. It has to be emphasized again that the $\hat{\beta}_i$ are the least squares estimators for the transformed model and not the original model. Having obtained $\hat{\beta}_i$ it will usually be necessary to transform back to the original model. This will lead to derived estimators of the parameters of this original model.

Example 1
Suppose the original model was

$$y = \alpha e^{\beta x} \cdot u$$

giving

$$\ln y = \ln \alpha + \beta x + e, \quad e = \ln u \cap N(0, \sigma^2)$$

then we can readily find the least squares estimator, $\widehat{\ln \alpha}$ of $\ln \alpha$, with estimated variance S_* say.
The natural estimator of α is then

$$\hat{\alpha}_* = \exp(\widehat{\ln \alpha})$$

If e is normally distributed with variance σ^2 then $\hat{\alpha}_*$ will have a lognormal distribution, which, using the properties given in Table 3.4 will have an estimated mean of $\hat{\alpha}_* e^{\frac{1}{2}s_*^2}$ and standard error of $\hat{\alpha}_* e^{\frac{1}{2}s_*^2}(e^{s_*^2} - 1)^{\frac{1}{2}}$. This suggests that a modified estimator $\hat{\alpha}_+ = \hat{\alpha}_* e^{-\frac{1}{2}s_*^2}$ might be more appropriate.

The previous example illustrates that even where a non-linear model can be linearized this does not remove all the problems associated with non-linearity. We have also seen that some non-linear models are just not amenable to linearization. The question now arises as to how to estimate the parameters in non-linear models. For complex non-linear models the methods available are mathematically involved and a range of sophisticated computer programs are available to implement the methods. However, in principle, the methods can be illustrated by a simple example. We illustrate three basic approaches. The first is based on searching in a systematic fashion for the parameter values that minimize the residual sum of squares. The second is to linearize the model using an approximation. The third is to use numerical techniques to solve the Normal Equations.

Example 2
We illustrate the three methods by considering the exponential model

$$y_i = \alpha e^{\beta x} + e_i$$

Notice that it is only the parameter β that causes the non-linearity. So if β is specified at any value, $\hat{\beta}$, we can apply least squares to the linear model and find an estimator of α. This we shall denote by $\hat{\alpha}$. We will assume that by graphical means, or using the method described in subsection 13.2.2, we have an initial estimate $\hat{\beta}$ of β.

(a) *Direct search*
If the estimator of β is taken as $\hat{\beta}$ then we define $z_i = e^{\hat{\beta}x_i}$ and write the model as

$$y_i - \alpha z_i + e_i.$$

This is the standard simple regression model for which the Least Squares Estimator is

$$\hat{\alpha} = \sum y_i z_i / \sum z_i^2$$

The residual for this fitted model is

$$\hat{e}_i = y_i - \hat{\alpha} e^{\hat{\beta}x_i} = y_i - \hat{\alpha} z_i$$

The residual sum of squares given any $\hat{\beta}$ will be

$$\text{RSS}(\hat{\beta}) = \sum \hat{e}_i^2 = \sum (y_i - \hat{\alpha} z_i)^2$$

Hence, given any value $\hat{\beta}$, we can straightforwardly calculate first $\hat{\alpha}$ and then $\text{RSS}(\hat{\beta})$. Hence we can undertake a systematic search to find the $\hat{\beta}$, and its associated $\hat{\alpha}$, that minimises the residual sum of squares. Table 13.5(c) shows a search for the data of Table 13.5(a). If such a process is to be carried out repeatedly, or with several non-linear parameters, the search procedure will have to be fairly elaborate and the other methods are probably more efficient. However, if the problem is a 'one-off' exercise, this method is simpe and effective and can be used whilst sitting at a computer and simply typing out values for $\hat{\beta}$ and getting $\text{RSS}(\hat{\beta})$ in return. It also has the merit of giving a feel for the sharpness of the minimum and thus the sensitivity of the estimators. A flat minimum means that small changes in the estimator values have little effect on the quality of the fit. Table 13.6 gives a further illustration of this approach for a situation involving two non-linear parameters.

(b) *Linearizing the model*
Though the term $e^{\beta x}$ is non-linear in β it can be approximated by a straight line provided β only varies by a small amount. This is illustrated in Figure 13.2(a), where the line AB approximates the curve AC If we are to use this result to change the non-linear model into an approximate linear model, then the approximation requires that the value $\hat{\beta}_0$, used as the basis for the approximation, must be close to the least squares value being searched for. It is seen from the figure that, for point C close to point B,

$$g(\beta) = g(\hat{\beta}_0) + (\beta - \hat{\beta}_0) m_0,$$

where m_0 is the slope at $\hat{\beta}_0$. For our exponential model $g(\beta) = \alpha e^{\beta x_i}$

so,
$$m_{0i} = \frac{d}{d\hat{\beta}_0}(\hat{\alpha}_0 e^{\hat{\beta}_0 x_i}) = \hat{\alpha}_0 x_i e^{\hat{\beta}_0 x_i} = \hat{\alpha}_0 x_i z_i,$$

using the notation of (a) and assuming that we have initial estimates $\hat{\alpha}_0, \hat{\beta}_0$. The non-linear model in β can thus be linearized as

$$y_i = \hat{\alpha}_0 e^{\beta x_i} + e_i$$
$$= \hat{\alpha}_0 e^{\hat{\beta}_0 x_i} + (\beta - \hat{\beta}_0) m_{0i} + e_i$$

Writing the term $\beta - \hat{\beta}_0$ as an adjustment, δ, this equation can be rewritten as

$$y_i - \hat{\alpha}_0 e^{\hat{\beta}_0 x_i} = \delta m_{0i} + e_i.$$

The left-hand side here is the residual, $\hat{e}_{0i}$, from the previous fitted model.

Table 13.5 Fitting a non-linear model

(a) Data: temperature measurements on a cooling body in ice
Model: Newtons law of cooling. Temp., $y_i = \alpha e^{\beta i}$, i = time
Measurement error independent of temperature suggesting

$$y_i = \alpha e^{\beta i} + e_i$$

Observations

i (mins)	1	2	3	4	5	6	7	8	9	10
y_i(°C)	96.9	88.4	79.0	70.5	65.1	60.9	54.5	50.3	46.2	41.8

(b) Initial Estimates. Method of 13.2.2.
$S_1 = 399.9$ $S_2 = 253.7$ $\hat{\beta} = -0.091$ $\hat{\alpha} = 104.2$

(c) Direct Search

$\hat{\beta}$	RSS($\hat{\beta}$)	$\hat{\alpha}$
−0.10	21.36	108.20
−0.09	12.84	103.94
−0.08	68.62	99.65
−0.091	10.85	104.37
−0.092	9.50	104.80
−0.093	8.79	105.23
−0.094	8.71	105.66
−0.095	9.26	106.08
−0.0932	8.72	105.31
−0.0934	8.68	107.40
−0.0936	8.67 →	105.48
−0.0938	8.68	105.57

Table 13.5 *(contd.)*

(d) Linearized model—a sequence of iterations

$\hat{\alpha}$	δ	$\hat{\beta}$
—	—	−0.1
108.2	0.00193	−0.0981
107.4	0.00134	−0.0967
106.8	0.00093	−0.0958
106.4	0.00065	−0.0952
106.1	0.00045	−0.0947
105.5	0.00000	−0.0936

(e) Using Normal Equations—a sequence of iterations starting with a poor $\hat{\beta}$

$\hat{\alpha}$	$\delta*$	$\hat{\beta}$
—	—	−0.2
146.7	0.06649	−0.1335
122.0	0.01534	−0.1182
115.8	0.00853	−0.1096
112.2	0.00527	−0.1044
110.0	0.00342	−0.1009
108.6	0.00228	−0.0986
105.5	0.00000	−0.0936

So the linearized model can be written as a regression of $\hat{e}$ on m thus

$$\hat{e}_{0i} = \delta m_{0i} + e_i.$$

The least squares estimator of δ in this linearized model is thus

$$\hat{\delta}_0 = \sum \hat{e}_{0i} m_{0i} / \sum m_{0i}^2,$$

using the standard result for a regression line through the origin, as in (a). Thus the revised estimator of β will be

$$\hat{\beta}_1 = \hat{\beta}_0 + \hat{\delta}_0.$$

We may now use $\hat{\beta}_1$ to find $\hat{\alpha}_1$ as in method (a) and start the process again. Thus the process gives an iterative procedure which, provided the approximation is valid, will lead to repeatedly improved estimators. The error in the linear approximation will depend on δ^2 and the rate of change of m, which is xm. A check on the sizes of $\delta^2 x_i m_i$ will indicate whether the approximation is likely to be satisfactory. These terms should be small compared with δm_i. Table 13.5(d) illustrates these calculations.

(c) *Solution of the normal equations*
The residual sum of squares is

$$S = \sum (y_i - \hat{\alpha} e^{\beta x_i})^2 = \sum (y_i - \hat{\alpha} z_i)^2$$

Table 13.6 A two-parameter search example

(a) The Model: The diabetes model given in section 9.2.
Assuming an administration of glucose to give a maximum at time zero the measured concentration of glucose in the blood at half hourly intervals is given by the solution form of the model which is

$$g_t = \alpha e^{-\beta t} \cos \frac{2\pi}{T} t + e_t \qquad t = 1, 2, 3, \ldots 21.$$

$T/2$ is the period of oscillation in hours.

(b) Data concentration relative to a norm, standardized units

hour	1	2	3	4	5	6	7
g	2.89	1.13	1.66	0.46	-2.98	-3.34	-2.36

hour	8	9	10	11	12	13	14
g	-2.57	2.39	1.66	0.94	3.46	-1.08	-1.35

hour	15	16	17	18	19	20	21
g	-1.25	-1.16	-1.72	-2.73	0.33	0.35	1.99

(c) Search β, T fixed, α estimated by Least Squares and Residual Sum of Squares RSS (β, T) calculated.
(i) RSS (β, T)

		T		
	6	8	10	12
0.01	71	77	27	75
0.05	70	75	26	57
β 0.10	71	72	27	56
0.15	72	70	40	55
0.20	73	69	48	57

(ii) RSS (β, T)

	T		
	9.5	10.0	10.5
0.020	35.68	26.40	26.61
0.025	35.22	26.10	26.49
0.030	34.83	26.00	26.52
β 0.035	34.54	25.95	26.61
0.040	34.32	26.00	26.71
0.045	34.20	26.13	27.02
0.05	34.1	26.34	27.21

Estimates at this stage
$\hat{\alpha} = 3.2$, $\hat{\beta} = 0.035$, $T = 10$

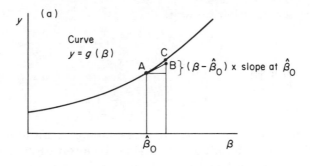

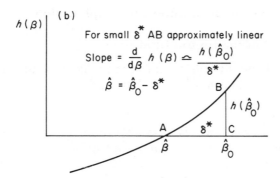

Figure 13.2 Linear approximations

Differentiating with respect to $\hat{\alpha}$ and equating to zero gives directly

$$\hat{\alpha} = \sum y_i z_i / \sum z_i^2.$$

Differentiating with respect to $\hat{\beta}$ and equating to zero gives

$$\sum x_i e^{\beta x_i}(y_i - \hat{\alpha}e^{\beta x_i}) = 0$$

The problem here is to find where this non-linear function of $\hat{\beta}$ is zero. Denoting the function by $h(\hat{\beta})$ it will be seen from Figure 13.2(b) that, if $h'(\hat{\beta})$ is the slope of $h(\hat{\beta})$, then the adjustment, δ^*, to $\hat{\beta}_0$ to give the zero value of $h(\beta)$ will be given approximately by

$$\delta^* = h(\hat{\beta}_0)/h'(\hat{\beta}_0).$$

Thus $\hat{\beta} = \hat{\beta}_0 - \delta^*$, provided δ^* is sufficiently small for the curve $h(\beta)$ to be treated as approximately linear in the region of the 'triangle' of Figure 13.2(b).
    Now

$$h(\hat{\beta}_0) = \sum x_i y_i e^{\hat{\beta}_0 x_i} - \hat{\alpha}_0 \sum x_i e^{2\hat{\beta}_0 x_i}$$

so

$$h'(\hat{\beta}_0) = \sum x_i^2 y_i e^{\hat{\beta}_0 x_i} - 2\hat{\alpha}_0 \sum x_i^2 e^{2\hat{\beta}_0 x_i}$$

Simplifying the notation we write

$$u_i = x_i y_i \, e^{\beta_0 x_i} = x_i y_i z_i$$

$$v_i = x_i \, e^{2\beta_0 x_i} = x_i z_i^2$$

Then

$$\delta^* = \frac{\sum u_i - \hat{\alpha}_1 \sum v_i}{\sum x_i u_i - 2\hat{\alpha} \sum x_i v_i}$$

Thus the calculation of δ^* is a matter of replacing an initial set of x_i, y_i by a number of new variables, z_i, u_i, v_i, using the latest estimates of α and β. From δ^* the estimate β can be revised and the whole process repeated. Table 13.5(e) illustrates.

13.4 Methods of maximum explanation

13.4.1 The method of maximum likelihood

This method of fitting distributional models is best understood by considering first a simple discrete distribution.

Example 1
Suppose a single observation r is available on a Binomial distribution $\text{Bin}(n, \theta)$. The probabilty of getting this observation is

$$\binom{n}{r} \theta^r (1 - \theta)^{n-r}.$$

This probability depends on θ through the likelihood

$$\mathscr{L}(\theta) = \theta^r (1 - \theta)^{n-r}. \qquad 0 \leqslant \theta \leqslant 1$$

which is sketched in Figure 13.3. It will be seen that the probability has a

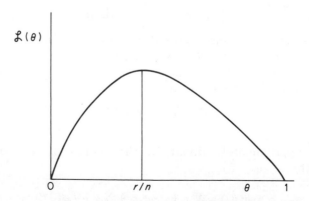

Figure 13.3 The binomial likelihoods

maximum at a value $\hat{\theta} = r/n$. So the observation r is more likely for this value than for any other value of θ. The function $\mathscr{L}(\theta)$ is the likelihood and $\hat{\theta}$ the *Maximum Likelihood Estimator* of θ. Remember that the shape of the likelihood is not affected by any multiplying terms, such as $\binom{n}{r}$, which do not depend on the parameter, so it is usual to leave them out for estimation problems.

Suppose now that there are m independent observations $r_1, r_2, \ldots r_m$, on the above Binomial situation. By the multiplication rule of probability the probability of this set of observations is

$$\binom{n}{r_1} \theta^{r_1}(1-\theta)^{n-r_1} \times \binom{n}{r_2} \theta^{r_2}(1-\theta)^{n-r_2} \times \ldots \times \binom{n}{r_m} \theta^{r_m}(1-\theta)^{n-r_n}$$

Ignoring the constants and expressing as simply as possible in terms of θ we get

$$\mathscr{L}(\theta) = \theta^{\Sigma r_i}(1-\theta)^{mn - \Sigma r_i}, \qquad 0 \leqslant \theta \leqslant 1.$$

If we write $r' = \Sigma r_i$ and $n' = mn$ it is seen that the form of the likelihood is as in the single observation situation, so the maximum likelihood estimator is

$$\hat{\theta} = r'/n' = \sum_{i=1}^{m} r_i/mn.$$

It is worth noting that the form of the likelihood reveals that, if r' and n' are known, knowing that n' is composed of mns and that the original observations were $r_1, r_2, \ldots, r_m$ is not relevant information. However, it is irrelevant only in so far as we assume the validity of the model. If there is a possibility that θ is not constant over the series of observations then this will only be detected by studying the separate estimators

$$\hat{\theta}_1 = r_1/n, \hat{\theta}_2 = r_2/n, \ldots, \hat{\theta}_m = r_m/n.$$

If a likelihood can be expressed in terms of some function of the observations, such as $r' = \Sigma r$ in this example, then the function is referred to as a *Sufficient Statistic*. Knowledge of the sufficient statistic is sufficient for the likelihood to be evaluated without knowledge of the original data.

Generalizing the above example we can state the method of maximum likelihood for discrete distributions is as follows:

If independent observations $r_1, r_2, \ldots, r_m$ are obtained from a situation modelled by a discrete distribution $p(r; \theta)$ then we calculate the likelihood as

$$\mathscr{L}(\theta) = p(r_1; \theta)p(r_2; \theta) \ldots p(r_m; \theta).$$

The *Maximum Likelihood Estimator* (MLE) of θ is the value, $\hat{\theta}$, that maximizes $\mathscr{L}(\theta)$. If there are several parameters $\theta_1, \theta_2, \ldots, \theta_k$, their ML estimators are the values $\hat{\theta}_1, \hat{\theta}_2, \ldots, \hat{\theta}_k$ that jointly maximize $\mathscr{L}(\theta_1, \theta_2,$

$\ldots, \theta_k)$. Thus ML estimators are the parameter values that maximize the probability of what we know to have happened, i.e. of the actual data.

Example 2

In an experiment on 'learning' a series of trials are carried out on five successive days $i = 1, 2, 3, 4, 5$. On each day a subject seeks to perform a task in a fixed time. He repeats until his first success. The first success occurred on trials 5, 7, 4, 2, 4. The assumed model is that the probability of failure decreases each day according to

$$Pr(\text{failure}|\text{day } i) = \alpha^i \qquad 0 < \alpha < 1$$

Thus the probability of first success at trial r on day i is

$$(\alpha^i)^{r-1}(1-\alpha^i).$$

The likelihood is therefore

$$\mathscr{L}(\alpha) = \alpha^4(1-\alpha)\cdot(\alpha^2)^6(1-\alpha^2)\cdot(\alpha^3)^3(1-\alpha^3)\cdot(\alpha^4)(1-\alpha^4)\cdot(\alpha^5)^3(1-\alpha^5)$$
$$= \alpha^{44}(1-\alpha)(1-\alpha^2)(1-\alpha^3)(1-\alpha^4)(1-\alpha^5)$$

To find the maximum of this analytically is rather fearsome. It is, however, a very simple formula to evaluate numerically on a programmable calculator or a computer. Thus values of α can be inserted and adjusted to find the maximum of $\mathscr{L}(\alpha)$. A value of approximately $\hat{\alpha} = 0.907$ is obtained through such a search.

Consider now the case of a parameter, θ, from a continuous distribution $f(x; \theta)$. In this case we cannot have the probability of an observation, x, but the analogous quantity is the p.d.f., $f(x; \theta)$, so we construct the likelihood of a set of independent observations $x_1, x_2, \ldots, x_n$ as

$$\mathscr{L}(\theta) = f(x_1; \theta) f(x_2; \theta) \ldots f(x_n; \theta).$$

The Maximum Likelihood Estimator is again the value of θ that gives for the actual data a larger likelihood than any other value of θ. It is often convenient to find the maximum of the log likelihood $L(\theta)$, where

$$L(\theta) = \ln \mathscr{L}(\theta),$$

for $L(\theta)$ is often easier to manipulate and the two curves have maxima at the same value of θ.

Example 3

Given observations $x_1, x_2, \ldots, x_n$ from the exponential distribution

$$f(x; \lambda) = \lambda e^{-\lambda x} \qquad x \geqslant 0, \lambda > 0,$$

then

$$\mathscr{L}(\lambda) = \lambda^n e^{-\lambda \sum_i^n x_i}$$

Figure 13.4 shows the form of $\mathscr{L}(\lambda)$.

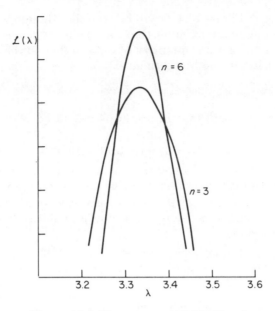

Figure 13.4 The exponential likelihood

Taking logs gives

$$L(\lambda) = n \ln \lambda - \lambda \sum_{1}^{n} x_i.$$

Differentiating with respect to λ gives

$$\frac{dL(\lambda)}{d\lambda} = \frac{n}{\lambda} - \sum x_i.$$

At the maximum, where $\lambda = \hat{\lambda}$, $dL(\lambda)/d\lambda$ which is the slope of $L(\lambda)$, is zero so

$$\frac{n}{\hat{\lambda}} - \sum x_i = 0$$

so the MLE is

$$\hat{\lambda} = n \bigg/ \sum_{1}^{n} x_i = 1/\bar{x}.$$

From the shape of Figure 13.4 it is clear that there will only be one point where $dL(\lambda)/d\lambda = 0$, so this is the maximum without further study. Note that in this case the MLE is the same as the method of moments estimator. Suppose we have two sets of observations

(a) 1.0, 2.5, 5.5,
(b) 0.5, 1.0, 1.4, 2.5, 5.5, 7.1,

simulated to give $\hat{\lambda} = 1/3$ for both sets of data. The likelihood for both (a) and (b) is plotted in Figure 13.4. It will be seen that though both likelihood functions have the same maximum, that for (b) is more peaked than that for (a). This reflects the greater quantity of data in (b) and the consequently increased precision in the ML estimate.

Example 4
The percentage of a pollutant in a lake is measured with samples from seven points

$$0.12, 0.15, 0.1, 0.15, 0.14, 0.16, 0.18$$

The simplest model for the percentage is a normal distribution, $N(\mu, \sigma^2)$. Though the percentage cannot be negative (or greater than 100) the model will be reasonable for most purposes, provided μ is greater than 4σ so that negative values from the model have very low probability. The p.d.f. for the normal distribution is

$$f(x; \mu, \sigma^2) = \frac{1}{\sqrt{2\pi}\,\sigma} \exp -\tfrac{1}{2}(x - \mu)^2/\sigma^2, \qquad -\infty < x < \infty.$$

The likelihood $\mathscr{L}(\mu, \sigma^2)$, for independent observations $x_1, \ldots, x_n$ reduces to

$$\mathscr{L}(\mu, \sigma^2) = \frac{1}{\sigma^n} \exp -\frac{1}{2\sigma^2} \sum (x_i - \mu)^2.$$

Taking logs of this gives

$$L(\mu, \sigma^2) = -n \ln \sigma - \frac{1}{2\sigma^2} \sum (x_i - \mu)^2.$$

Differentiating with respect to μ gives

$$\frac{\partial L}{\partial \mu} = \frac{2}{2\sigma^2} \sum (x_i - \mu).$$

Equating to zero for maximum gives

$$\sum (x_i - \hat{\mu}) = 0,$$

so

$$\hat{\mu} = \sum_{i=1}^{n} x_i/n = \bar{x},$$

which gives for the data $\hat{\mu} = 0.151$.

The method of maximum likelihood gives the same answer as the method of moments. To find the value of σ which maximizes $L(\mu, \sigma^2)$ we differentiate with respect to σ giving

$$\frac{\partial L}{\partial \sigma} = -\frac{n}{\sigma} + \frac{2}{2\sigma^3} \sum (x_i - \mu)^2.$$

Equating this to zero and setting $\mu = \hat{\mu} = \bar{x}$ to get the overall maxima we have

$$\hat{\sigma}^2 = \frac{1}{n}\sum(x_i - \bar{x})^2 = s'^2$$

so for the data $\hat{\sigma} = 0.009$.

To consider the shape of the likelihood function, note that

$$\sum(x_i - \mu)^2 = \sum[(x_i - \bar{x}) - (\mu - \bar{x})]^2$$
$$= \sum(x_i - \bar{x})^2 + n(\bar{x} - \mu)^2,$$

as $\qquad \Sigma(x_i - \bar{x}) = 0.$

Thus

$$\mathscr{L}(\mu, \sigma^2) = \frac{1}{\sigma^n}\exp-\frac{1}{2\sigma^2}\sum(x_i - \bar{x})^2 \cdot \exp-\frac{1}{2\sigma^2}n(\bar{x} - \mu)^2$$

$$= \exp-\frac{ns'^2}{2\sigma^2} \cdot \frac{1}{\sigma^n}\exp-\frac{n}{2\sigma^2}(\mu - \bar{x})^2.$$

The shape of this as a function of μ, is the shape of a normal curve $N(\bar{x}, \sigma^2/n)$, which for the data used is $N(0.151, (0.0034)^2)$. Thus $\bar{x}$ is the maximum of μ and σ^2/n gives a measure of the precision of this estimator, since it determines the sharpness of the peak around $\hat{\mu} = \bar{x}$. This corresponds to the previous argument that the precision of $\bar{x}$ is indicated by its standard error (deviation) which is $\sigma/\sqrt{n}$.

Example 5
Consider again the regression model

$$y_i = \beta_0 + \beta_1 x_i + e_i$$

and assume that $e_i \cap N(0, \sigma^2)$. The likelihood for the e_i, using the results of the last example is

$$\frac{1}{\sigma^n}\exp-\frac{1}{2\sigma^2}\sum e_i^2$$

so $\qquad L = -n\ln\sigma - \frac{1}{2\sigma^2}\sum e_i^2$

$$= -n\ln\sigma - \frac{1}{2\sigma^2}\sum(y_i - \beta_0 - \beta_1 x_i)^2.$$

Thus the maximum likelihood estimators $\hat{\beta}_2$, $\hat{\beta}_1$ are the values that minimize (note the $-$ sign of $-1/2\sigma^2$)

$$\sum(y_1 - \hat{\beta}_0 - \hat{\beta}_i x_i)^2,$$

which is $\Sigma\hat{e}_i^2$. This is the criterion that was used for the method of Least Squares so the two methods are equivalent, for Normal e, for estimating regression coefficients. Maximum likelihood does, however, provide an estimate of σ. For

$$L = -n \ln \sigma - \frac{1}{2\sigma^2} \sum \hat{e}_i^2$$

so

$$\frac{\partial L}{\partial \sigma} = -\frac{n}{\sigma} + \frac{1}{\sigma^3} \sum \hat{e}_i^2$$

equating to zero for maximum gives

$$\hat{\sigma}^2 = \frac{1}{n} \sum \hat{e}_i^2 = \frac{1}{n} [\text{Residual Sum of Squares}].$$

Example 6

Consider a second order autoregressive model

$$y_t = \phi_1 y_{t-1} + \phi_2 y_{t-2} + e_t,$$

where $e_t \cap N(0, \sigma^2)$. We can derive the likelihood function in an identical fashion to the last example, provided we can make some allowance for the 'start-up' of the model, since y_1 and y_2 formally depend on non-existent y_0 and y_{-1}. The simplest solution to this problem is to take y_1 and y_2 as constants and regard the model as applying only for $t = 3, 4, \ldots$. Hence the log likelihood will be

$$L = -(n-2) \ln \sigma - \frac{1}{2\sigma^2} \sum_{t=3}^{n} (y_t - \phi_1 y_{t-1} - \phi_2 y_{t-2})^2.$$

So the estimation reduces to a regression of y_t on y_{t-1} and y_{t-2}. If there is little data, we cannot ignore the evidence in y_1 and y_2, then a more complex likelihood is involved and numerical methods are needed to solve the equations.

Example 7

As a final illustration of the direct application of Maximum Likelihood we derive the estimators for the parameters of the Poisson and Geometric distributions, already used in section 12.6. The calculations are laid out to emphasize the structure of the method. The data is symbolized as $r_1, r_2, \ldots, r_n$.

Calculation	*Poisson*	*Geometric*
The p.d.f. $P(r\|\theta)$ $(\theta = \mu, \phi)$	$e^{-\mu} \dfrac{\mu^r}{r!}$	$(1-\phi)\,\phi^r$
Initial Likelihood	$\displaystyle\prod_{i=1}^{n} e^{-\mu} \frac{\mu^{r_i}}{r_i!}$	$\displaystyle\prod_{i=1}^{n} (1-\phi)\,\phi^{r_i}$
Likelihood $\mathscr{L}(\theta)$	$e^{-n\mu}\,\mu^{\Sigma r_i}$	$(1-\phi)^n\,\phi^{\Sigma r_i}$

Log Likelihood $L(\theta)$	$-n\mu + \sum r_i \ln \mu$	$n \ln(1-\phi) + \sum r_i \ln \phi$
$\dfrac{\mathrm{d}L}{\mathrm{d}\theta}$	$-n + \dfrac{\sum r_i}{\mu}$	$\dfrac{-n}{1-\phi} + \dfrac{\sum r_i}{\phi}$
$\dfrac{\mathrm{d}L}{\mathrm{d}\theta} = 0 \qquad \hat{\theta} = .$	$\hat{\mu} = \dfrac{\sum r_i}{n}$	$\hat{\phi} = \dfrac{\sum r_i}{n + \sum r_i}$

A useful property of methods based on maximization or minimization arises when interest focusses not on the natural parameter, θ, but on some increasing or decreasing (monotone) function, $g(\theta)$, of it. A standard mathematical result is that the value of $g(\theta)$ leading to maximum likelihood (or least squares) is just $g(\hat{\theta})$ where $\hat{\theta}$ is the maximum likelihood (or least squares) estimator of θ.

Example 8
If the time to failure of a component is exponentially distributed the probability of being operational at time T, its reliability at T, is given by

$$R_T(\lambda) = Pr(t \geqslant T),$$

which is

$$R = R_T(\lambda) = e^{-\lambda T}.$$

If data on failure times $t_1, \ldots, t_n$ is available the likeihood is

$$\mathscr{L}(\lambda) = \lambda^n e^{-\lambda \sum t_i}$$

which in terms of R is

$$\mathscr{L}(R) = \left(-\frac{1}{T} \ln R \right)^n R^{\sum t_i / T}.$$

We could find the maximum likelihood estimator $\hat{R}$ of R directly from this, but the result just stated provides an easier approach for, by using the ML estimator ($\hat{\lambda} = n/\sum t_i$), we can write

$$\hat{R} = R(\hat{\lambda}) = e^{-\hat{\lambda} T} = e^{-nT/\sum t_i},$$

as the ML estimator of the reliability at time T.

Note that this form of result does not apply to estimators obtained by shape equalization methods, since the moments of a function of a random variable are not equal to the function of the moments. The order of the operations involved cannot be reversed, as they can be for maximization and minimization of monotone functions.

As a final example of the use of maximum likelihood we turn to an example that illustrates the use of numerical procedures to solve the more awkward equations often produced by using the method.

Example 9*
Consider again Example 4 of Section 13.2.2, which involves the Poisson distribution with no zero observation, namely

$$Pr(R = r) = e^{-\mu}\frac{\mu^r}{r!} \bigg/ (1 - e^{-\mu}), \ r = 1, 2, 3, \ldots .$$

The relevant data is given in Table 19.2. We denote the total number of households by N and the number of people in the ith household by r_i. The likelihood is

$$\mathscr{L}(\mu) = \prod_{i=1}^{N}\left[\frac{e^{-\mu}}{1 - e^{-\mu}}\frac{\mu^{r_i}}{r_i!}\right].$$

Noting that each r occurs with a frequency f_r in a table such as Table 19.2, this can be rewritten as

$$\mathscr{L}(\mu) = \left(\frac{e^{-\mu}}{1 - e^{-\mu}}\right)^N \cdot \mu^{\Sigma r f_r},$$

dropping the constant term $(r_i!)$. From Table 19.2, $N = 592$ and $\Sigma r f_r = 779$, which we will denote by T. As usual it is somewhat easier to work with the log likelihood, $L(\mu)$, which is

$$L(\mu) = -N\mu - N\ln(1 - e^{-\mu}) + T\ln\mu.$$

Differentiating with respect to μ gives a quantity, $U(\mu)$, which is often called the *Score*,

$$\frac{dL}{d\mu} = U(\mu) = -N - \frac{Ne^{-\mu}}{1 - e^{-\mu}} + \frac{T}{\mu} = \frac{-N}{1 - e^{-\mu}} + \frac{T}{\mu}.$$

Equating the score to zero to obtain the maximum likelihood estimate $\hat{\mu}$ leads to the relation

$$\frac{\hat{\mu}}{1 - e^{-\hat{\mu}}} = \frac{T}{N}.$$

Notice that the form of the left-hand side is that of the expectation $E(r)$ and that of the right-hand side is simply the average of the observed r. Thus the result is the same expression as one would derive from the method of moments. Unfortunately the form of this equation does not lend itself to a simple explicit solution. We thus need to make use of an iterative solution method. To do this we need a first approximation. We obtained this in Example 4 of subsection 13.2.2 using the analogy approach to estimation. The value obtained was

$$\hat{\mu}_0 = 0.610.$$

We now need a general procedure for taking an estimate and improving upon it. Note first that in terms of the score the maximum likelihood estimate, $\hat{\mu}$, satisfies the equation

$$U(\hat{\mu}) = 0.$$

If $\hat{\mu}_0$ is reasonably close to $\hat{\mu}$ then the curve of $U(\hat{\mu})$ can be replaced by a straight line approximation as was done in solving the normal equations in sub-section 13.3.3. This is illustrated in Figure 13.5. The slope of this line can be obtained:

(a) by calculus as the derivative of $U(\hat{\mu})$, at $\hat{\mu} = \hat{\mu}_0$. This derivative $dU/d\mu$ is often denoted by $-V(\mu)$;
(b) from the triangle as

$$\frac{U(\hat{\mu}_0)}{\hat{\mu}_0 - \hat{\mu}_1}.$$

Equating these two gives

$$\hat{\mu}_1 = \hat{\mu}_0 + \frac{U(\hat{\mu}_0)}{V(\hat{\mu}_0)}.$$

This is called the Newton–Raphson method, which we have already seen in relation to Least Squares estimation. Provided the first approximation is a reasonable one, $\hat{\mu}_1$ will be closer to $\hat{\mu}$ than was $\hat{\mu}_0$. The process can then be iterated to get better and better approximations.

To apply this method to our problem we first need the derivative of $-U(\hat{\mu})$,

$$V(\hat{\mu}) = -\frac{N\,e^{-\hat{\mu}}}{(1 - e^{-\hat{\mu}})^2} + \frac{T}{\hat{\mu}^2}.$$

Substituting for $\hat{\mu}_0$ in the above expressions gives

$$U(\hat{\mu}_0) = -19.35$$
$$V(\hat{\mu}_0) = 1638.12$$

so

$$\hat{\mu}_1 = \hat{\mu}_0 + \frac{U(\hat{\mu}_0)}{V(\hat{\mu}_0)} = 0.610 - 0.010 = 0.600.$$

The next iteration gives a negligible value of $U(\hat{\mu}_0)/V(\hat{\mu}_0)$ so the process has given a satisfactory answer to three significant figures on the first

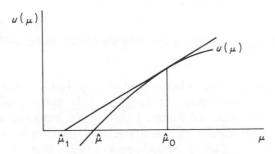

Figure 13.5 Approximating the score

iteration. Notice that the fact that we obtained a good approximation at the first iteration, depended on the quality of the first estimate, which in turn depended on the fact that in the data itself most observations fell in the first two cells of the table.

13.4.2 Maximum multiple correlation

It was suggested when discussing the fit-observation diagram, which plotted $\hat{y}$ versus y, that the multiple correlation coefficient $R^2 = r_{y\hat{y}}^2$ was a convenient measure of the quality of fit. It has the merit compared with the residual sum of squares, $\Sigma \hat{e}_i^2$, that its numerical value, lying in $(0, 1)$, has an intuitive interpretability, a value $r_{y\hat{y}}^2 = 0.95$ suggests immediately a very good fit. For the linear model the two quantities are directly related by

$$r_{y\hat{y}}^2 = 1 - \frac{\sum \hat{e}_i^2}{\sum (y_i - \bar{y})^2}.$$

It follows that fitting by minimizing $\sum \hat{e}_i^2$ leads directly to the fitted model with the maximum multiple correlation.

The multiple-correlation coefficient is often used as the standard measure of fit for models. With very large models the evaluation of the fitted values, $\hat{y}$, for any given set of parameter values may be a significant job. In such cases an exploratory approach has often been taken, using prior information to the full, to seek a set of parameter values that maximise $r_{y\hat{y}}^2$. Here $r_{y\hat{y}}^2$ is seen as the measure of the ability of the model, represented by $\hat{y}$, to explain the data, y. Thus maximum $r_{y\hat{y}}^2$ can be seen as another method of 'maximum explanation'. It is, however, a method that needs handling with care. To see why, we need to consider the case of non-linear models. For non-linear models the relation between $r_{y\hat{y}}^2$ and $\sum \hat{e}_i^2$ ceases to hold, so that they become different criteria. The model with the maximum $r_{y\hat{y}}^2$ may not be the one with the minimum $\sum \hat{e}_i^2$. The difference between the two and the danger of using only $r_{y\hat{y}}^2$ is illustrated in Figure 13.6.

In Figure 13.6(a) it is seen that the $45°$ line gives the line of perfect fit. The 'vertical' distances from this line give the residuals $\hat{e}$. The problem of $r_{y\hat{y}}^2$ is that it is only a measure of linearity. Thus the points in Figure 13.6(b), where there is a systematic over-estmation of y for small $\hat{y}$ and vice versa, also give $r_{y\hat{y}}^2 \simeq 1$. The value $r_{y\hat{y}}^2 = 1$ shows, not that $y = \hat{y}$ but, that:

$$y = \text{constant} + \text{constant } \hat{y},$$

i.e. that there is a linear relation between y and $\hat{y}$. Thus the model or parameter values leading to the largest $r_{y\hat{y}}^2$ is not necessarily the best fitting model. The form of Figure 13.6(b) does however suggest that a simple linear modification of the fitted model will give a good fit. The obvious ways to avoid being misled are to always plot the fit-observation diagram and to evaluate $\sum \hat{e}_i^2$ alongside $r_{y\hat{y}}^2$.

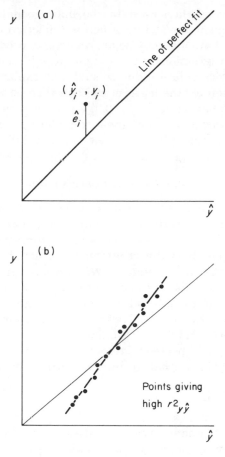

Figure 13.6 The interpretation of high $r_{y\hat{y}}^2$

The use of $r_{y\hat{y}}^2$ for progression models needs particular caution. If y_t is an economic series that is modelled using other economic variables, then it is likely that all the variables will reflect something of the general trends and swings shown by an economy. This will automatically lead to a high multiple correlation between y and $\hat{y}$, irrespective of any real causal relation between the variables. There are two approaches in this situation. One is to model all variables separately with trends and oscillations and subtract these effects from the data giving detrended and/or deseasonalized data. For example, if the fitted trend model is

$$y_t = 0.3 + 0.8t,$$

then the adjusted data would be

$$y'_t = y_t - 0.3 - 0.8\,t.$$

The adjusted data is then used in the main model and the multiple correlation coefficient then gives a better reflection of the explanatory power of the model. An alternative approach is to produce a comparative model for y using regression on trending and cyclic terms only. This then gives the multiple correlation due to such non-explanatory factors. The multiple correlation for the main model can then be compared with this standard model. For example if $r^2_{y\hat{y}} = 0.91$ for the main model this suggests a good model. If, however, a simple linear trend for y gave $r^2_{y\hat{y}} = 0.88$ then the quality of the main model has to be viewed in a different light.

13.5 Interval estimation

In our study of estimation we have been concerned with obtaining a single numerical value for each unknown parameter, that is to say with the problems of *Point Estimation*. In some circumstances it may be valuable to obtain a range of possible values that might include the parameter. This is the problem of *Interval Estimation*. We can use the data to obtain an interval (θ_1, θ_2) that is likely to contain the true value θ. Clearly on any given occasion such an interval will or it will not contain the true θ. Therefore we cannot in an objective sense talk of the probability of θ lying in the interval. As θ_1 and θ_2 are calculated from the data the interval is a random one. Hence in repeated trials we can obtain different intervals and calculate the probability that an interval surrounds the true value.

Example 1
Consider a set of observations $x_1, x_2, \ldots, x_n$ on a normal distribution $N(\mu, \sigma^2)$ where σ^2 is known. We have already noted two relevant points (a) that for a normal sample, means $\bar{x}$ have the distribution $N(\mu, \sigma^2/n)$, (b) that 95% of observations on a normal distribution $N(\mu, \sigma^2)$ lie in the interval $(\mu - 1.96\sigma, \mu + 1.96\sigma)$. From these two observations we can write

$$Pr\left(\mu - 1.96 \frac{\sigma}{\sqrt{n}} \leqslant \bar{x} \leqslant \mu + 1.96 \frac{\sigma}{\sqrt{n}} \right) = 0.95.$$

Re-writing the inequality we have

$$Pr\left(\bar{x} - 1.96 \frac{\sigma}{\sqrt{n}} \leqslant \mu \leqslant \bar{x} + 1.96 \frac{\sigma}{\sqrt{n}} \right) = 0.95.$$

Thus the probability that the random interval

$$\left(\bar{x} - 1.96 \frac{\sigma}{\sqrt{n}}, \bar{x} + 1.96 \frac{\sigma}{\sqrt{n}} \right)$$

surrounds the true mean μ is 0.95. Thus, if this calculation were repeatedly used, approximately 95% of the intervals so calculated would contain μ. Such an interval we call the 95% *Confidence Interval*. For the normal data of Example 4 of Section 13.4 the 95% confidence interval for μ is $(0.151 \pm 1.96 \times 0.0034)$ which is $(0.144, 0.158)$. Notice that in obtaining this interval we chose, on intuitive grounds, the distribution of the standard estimator, $\bar{x}$, of μ, as the basis for the calculation and that we chose a symmetrical interval about μ, though we could have chosen an infinity of other intervals containing 95% of the distribution. As the normal distribution is symmetric it follows that the width of a symmetric interval is smaller than that of a non-symmetric interval, e.g. the 95% interval between, for example, $\bar{x} - 2.58\left(\dfrac{\sigma}{\sqrt{n}}\right)$ and $\bar{x} + 1.70\left(\dfrac{\sigma}{\sqrt{n}}\right)$. It is reasonable to seek intervals that are as short as possible, though we might on occasions prefer those with equal probabilities at each end. These two will only be identical for symmetrical distributions.

When we seek to obtain some idea of the standard error of an estimate or to create a confidence interval for a parameter, we may sometimes be faced with the problem that the variance of the estimator depends on the unknown parameter value. This occurs in a number of common situations:

Binomil Bin(n, θ) data r

$$\hat{\theta} = r/n \qquad\qquad V(\hat{\theta}) = \frac{\theta(1-\theta)}{n}$$

Poisson P(μ) data $r_1, \ldots, r_n$

$$\hat{\mu} = \bar{r} \qquad\qquad V(\hat{\mu}) = \frac{\mu}{n}$$

Clearly one way of dealing with the problem is to replace the parameter by its estimate, this will give an approximate answer with little indication of the quality of the approximation. An alternative approach is to use a *Variance Stabilizing Transformation*. For the above situations we have:

Binomial Bin(n, θ) data r

$$z = 2\sin^{-1}\frac{r}{\sqrt{n}} \qquad \text{is approximately normally distributed}$$

$$E(z) = 2\sin^{-1}\sqrt{\theta} \qquad V(z) \simeq \frac{1}{n}$$

Poisson P(μ) data $r_1, \ldots, r_k$

$$z = 2\sqrt{\bar{r}} \qquad\qquad \text{is approximately normally distributed}$$

$$E(z) = 2\sqrt{\mu} \qquad\qquad V(z) \simeq \frac{1}{n}$$

It will be seen that in each case the approximate variance of the transformed variable depends on n but not on the unknown parameter. This facilitates the formulation of confidence intervals.

Example 2

Given 10 observations from a Poisson situation $P(\mu)$,

$$3, 2, 6, 0, 1, 3, 3, 8, 4, 2$$

$\hat{\mu} = \bar{r} = 3.2$ so the approximate standard error of $\hat{\mu}$, using $V(\bar{r}) = \hat{\mu}/n$, is $3.2/\sqrt{10} = 0.566$. We could thus give approximate 95 % confidence limits for μ as $(3.2 \pm 1.96 \times 0.566) = (2.1, 4.3)$, assuming $\bar{r}$ is approximately Normal, by the Central Limit Theorem.

Alternatively making use of the transformation $2\sqrt{\bar{r}}$ is approximately $N(2\sqrt{\mu}, 1/10)$, hence the 95 % limits are

$$2\sqrt{3.2} - 1.96 \frac{1}{\sqrt{10}} < 2\sqrt{\mu} < 2\sqrt{3.2} + 1.96 \frac{1}{\sqrt{10}}$$

$$1.48 < \sqrt{\mu} < 2.10$$

so

$$2.19 < \mu < 4.41$$

The second approximation reflects the skewness of the Poisson distribution.

Chapter 14

Conceptual aspects of estimation

14.1 Introduction

Conceptual aspects of fitting arise when we consider the properties we desire of estimators and of fitted models. The most common approach to defining such properties is in terms of the sampling distributions of estimators and of the moments of these distributions. The use of sampling distributions focusses attention on the properties that the estimators will have, over repeated application of the methods to sets of data from the given model. A consequence of this approach is that we need to make a change in notation for this chapter. In the past we have used $x_1, x_2, \ldots, x_n$ to denote the actual numbers. In this chapter we treat the $x_1, \ldots, x_n$ as a potential sample from the population of samples and use quantities like $E(\Sigma x_i)$ and $V(\Sigma x_i)$ with the x_i regarded as random variables from the distributions indicated by the model.

14.2 Biased and unbiased estimators

An estimator $\hat{\theta}$ of θ is said to be *unbiased* if $E(\hat{\theta}) = \theta$.

If an estimator is not unbiased we can define its *bias* by

$$b(\theta) = E(\hat{\theta}) - \theta$$

If sometimes happens that $b(\theta)$ depends on the number of observations and approaches zero as n increases. In this case $\hat{\theta}$ is said to be *Asymptotically Unbiased*. Remembering that the expectation represents a long-run average, bias is seen as an average deviation of the estimator from the true value.

Example 1

Let R $\cap$ Binomial B(n, θ) with independent observations $r_1, r_2, \ldots, r_m$. It was shown in the last chapter that the maximum likelihood estimator is

$$\hat{\theta} = \sum r_i / nm.$$

Taking expectations

$$E(\hat{\theta}) = E(\sum r_i/nm)$$
$$= \sum E(r_i/nm).$$

For a Binomial observation the expectation is $n\theta$ so

$$E(\hat{\theta}) = \sum_{i=1}^{m} n\theta/nm = \theta.$$

Hence $\hat{\theta}$ is an unbiased estimator of θ.

Example 2
Consider the linear regression model

$$y_i = \theta x_i + e_i, \qquad e_i \cap (0, \sigma^2)$$

The least squares estimator is

$$\hat{\theta} = \sum x_i y_i / \sum x_i^2.$$

Now

$$E(\hat{\theta}) = E(\sum x_i y_i / \sum x_i^2) = \sum x_i E(y_i) / \sum x_i^2,$$

but $\qquad E(y_i) = \theta x_i \qquad$ so

$$E(\hat{\theta}) = \sum (x_i \cdot \theta x_i) / \sum x_i^2 = \theta,$$

$\hat{\theta}$ is thus an unbiased estimator.

Notice that we have not needed to know the actual distribution of e, only that it has zero expectation. An alternative estimator, might be

$$\theta^* = \sum y / \sum x,$$

The expectation here is

$$E(\hat{\theta}^*) = E(\sum y_i / \sum x)$$
$$= \sum \theta x_i / \sum x_i = \theta$$

so θ^* is also unbiased.

When considering bias, care needs to be taken with functions of parameters. In general if $g(\theta)$ is some function of a parameter and $E(\hat{\theta}) = \theta$,

$$E[g(\hat{\theta})] \neq g(\theta).$$

Example 3
For the exponential distribution

$$f(x) = \lambda e^{-\lambda x}$$

the maximum likelihood estimator of λ is $\hat{\lambda} = 1/\bar{x}$. As indicated above we cannot find

$$E\left(\frac{1}{\bar{x}}\right) \quad \text{from} \quad E(\bar{x})$$

in any simple fashion. The only way to tackle this problem is to find the distribution of $\bar{x}$, hence the distribution of $1/\bar{x}$ and then its expectation. This involves mathematics beyond the scope of this book but in this particular case $\hat{\lambda}$ is found to be asymptotically unbiased.

Example 4

If we estimate variability, the problem occurs that most intuitive estimators and maximum likelihood estimators are biased. In section 10.3 we introduced

$$s'^2 = \frac{1}{n}\sum d_i^2, \quad d_i = x_i - \bar{x}$$

as the sample variance corresponding to the population variance σ^2. It can be shown that

$$E(s'^2) = \frac{n-1}{n}\sigma^2.$$

Thus s'^2 is biased and an unbiased estimator is given by

$$s^2 = \frac{1}{n-1}\sum d_i^2.$$

Notice that the divisor $n-1$, the Degrees of Freedom, represents the number of independent observations available for estimating σ^2. Given any $n-1$ observations plus the value of $\bar{x}$, calculated preliminary to finding the d_i, the remaining observation can be reconstructed. This procedure for calculating an unbiased estimator generalizes. For a linear regression $y = \alpha + \beta x + e$, two estimated parameters, $\hat{\alpha}$, $\hat{\beta}$ are required before the residual $\hat{e}$ can be found, thus $n-2$ degrees of freedom remain for estimating σ_e^2. The unbiased estimator of σ_e^2 is

$$s^2 = \frac{1}{n-2}\sum \hat{e}^2.$$

14.3 Efficiency

If we have an unbiased estimator then a measure of the closeness of $\hat{\theta}$ to θ is the variance of $\hat{\theta}$

$$V(\hat{\theta}) = E[(\hat{\theta} - \theta)^2], \quad \text{as} \quad E(\hat{\theta}) = \theta.$$

The smaller the variance the better.

Example 1

Consider the two estimators of θ in the regression model for road distances

$$\hat{\theta} = \sum x_i y_i / \sum x_i^2$$

and

$$\hat{\theta}* = \sum y_i / \sum x_i$$

Calculating their variances gives

$$V(\hat{\theta}) = V(\sum x_i y_i / \sum x_i^2)$$
$$= \sum x_i^2 \, V(y_i)/(\sum x_i^2)^2$$
$$= \sigma^2/(\sum x_i^2)$$
$$V(\hat{\theta}^*) = V \sum y / \sum x$$
$$= \sum V(y)/(\sum x)^2$$
$$= n\sigma^2/(\sum x)^2$$

To compare these two variances we note first that

$$\sum(x_i - \bar{x})^2 = \sum x_i^2 - 2\bar{x}\sum x + n\bar{x}^2$$
$$= \sum x_i^2 - \frac{(\sum x_i)^2}{n}.$$

The left-hand side is positive so

$$\sum x_i^2 > \frac{(\sum x_i)^2}{n}$$

and

$$\frac{n}{(\sum x_i)^2} > \frac{1}{\sum x_i^2}$$

hence $V(\hat{\theta}^*) > V(\hat{\theta})$.

Thus the estimators $\hat{\theta}$ are distributed more tightly about θ than the estimators $\hat{\theta}^*$. It can in fact be proved that for linear regression models least squares estimators have smaller variances than any other unbiased estimators that are linear in the observations y_i. Using the data of Table 13.1 and the estimator $\hat{\sigma} = 2.37$, the estimated standard deviations of $\hat{\theta}$ and $\hat{\theta}^*$ are found by these formulae to be $\hat{\sigma}_{\hat{\theta}} = 0.030$ and $\hat{\sigma}_{\hat{\theta}*} = 0.033$.

Example 2

An alternative estimator to $\bar{x}$ of μ in a Normal distribution is the sample median, *Me*. The median can be shown to have expectation μ, thus *Me* is an unbiased estimator. Its variance is $(\pi/2)(\sigma^2/n)$. Comparing $\bar{x}$ and *Me* gives

$$\frac{V(\bar{x})}{V(Me)} = \frac{2}{\pi} = 0.6366$$

This is termed the *Relative Efficiency* of $\bar{x}$ and *Me*. As $\bar{x}$ is more efficient than *Me*, the question arises whether an even more efficient estimator might be found if one searched long enough. A result in statistical theory shows that for any parameter in a distribution it is possible to calculate a quantity $V_{\min}$ such that for any estimator $\hat{\theta}$

$$V(\hat{\theta}) \geqslant V_{\min}$$

Thus there is a minimum possible variance, V_{min}. For unbiased estimators of μ in the normal distribution it is found that $V_{min} = \sigma^2/n$ which is the variance of $\bar{x}$. Thus $\bar{x}$ has a variance which cannot be improved.

The relative efficiency formula shows that a median calculated with 32 observations gives the same variance as the mean calculated with 20 observations. However, a property of the median is that it is not unduly influenced by distortions in the tails of the distributions or by errors in some observations. It may well be that it is worth paying for twelve extra observations to obtain such a more robust estimator.

14.4 Mean square error

A measure of how close $\hat{\theta}$ is to θ is

$$\text{MSE}(\hat{\theta}) = E\,[(\hat{\theta} - \theta)^2\,]$$

This is called the *Mean Square Error*. A small value for $\text{MSE}(\hat{\theta})$ is intuitively a desirable feature for an estimator. To see how this relates to the variance and bias of $\hat{\theta}$ we develop the following argument

$$\text{MSE}(\hat{\theta}) = E\,[(\{\hat{\theta} - E(\hat{\theta})\} - \{\theta - E(\hat{\theta})\})^2\,]$$
$$= E\,[\{\hat{\theta} - E(\hat{\theta})\}^2\,] - 2E\{\hat{\theta} - E(\hat{\theta})\}\{\theta - E(\hat{\theta})\} + \{\theta - E(\hat{\theta})\}^2$$

The first term is the variance of $\hat{\theta}$, the second is zero $\{E(\hat{\theta}) - E(\hat{\theta})\}$, the third is the bias squared. Thus

$$\text{MSE}(\hat{\theta}) = V(\hat{\theta}) + b(\hat{\theta})^2,$$

and the MSE is seen to be a combination of our previous two main measures of estimator quality. It is seen that for unbiased estimators seeking a small mean square error is identical to seeking a small variance.

Example 1

For a uniform distribution, $X \cap U(0, \theta)$ the maximum likelihood estimator is the largest observation $x_{(n)}$, $\hat{\theta} = x_{(n)}$. It can be shown that

$$E(x_{(n)}) = \frac{n}{n+1}\,\theta \quad \text{and} \quad V(x_{(n)}) = \frac{n}{(n+1)(n+2)}\,\theta^2$$

These formulae can be used to construct the properties of the three estimators shown in Table 14.1. It will be seen that $\hat{\theta}_2$ is unbiased but has the largest variance of the three. $\hat{\theta}_1$ is biased but has a small variance. Estimator $\hat{\theta}_3$ has a smaller bias and a smaller variance than $\hat{\theta}_2$, these combine to achieve a smaller MSE than the other two estimators.

14.5 Consistency

A reasonable requirement of an estimator is that it tends towards the true value, as the number of observations increases to infinity. It can be shown

Table 14.1 Comparison of estimator properties

$\hat{\theta}$	$X \cap U(0, \theta)$ ordered data $x_{(1)}, x_{(2)}, \ldots, x_{(n)}$		
	$E(\hat{\theta})$	$V(\hat{\theta})$	$\text{MSE}(\hat{\theta})$
$\hat{\theta}_1 = x_{(n)}$	$\dfrac{n}{n+1}\theta$	$\dfrac{n}{(n+1)^2(n+2)}\theta^2$	$\dfrac{2}{(n+1)(n+2)}\theta^2$
$\hat{\theta}_2 = \dfrac{n+1}{n}x_{(n)}$	θ	$\dfrac{1}{n(n+2)}\theta^2$	$\dfrac{1}{n(n+2)}\theta^2$
$\hat{\theta}_3 = \dfrac{n+2}{n+1}x_{(n)}$	$\dfrac{n(n+2)}{(n+1)^2}\theta$	$\dfrac{n(n+2)}{(n+1)^4}\theta^2$	$\dfrac{1}{(n+1)^2}\theta^2$
Putting $n = 5$			
$\hat{\theta}_1$	$0.83\,\theta$	$0.0198\,\theta^2$	$0.0476\,\theta^2$
$\hat{\theta}_2$	θ	$0.0286\,\theta^2$	$0.0286\,\theta^2$
$\hat{\theta}_3$	$0.97\,\theta$	$0.0270\,\theta^2$	$0.0278\,\theta^2$

that this will happen if the estimator $\hat{\theta}$ is either unbiased or asymptotically unbiased and also has a variance that tends to zero as n increases.

Example 1
For a Normal distribution, $N(\mu, \sigma^2)$, the standard estimator of μ is the average $\bar{x}$, for which $E(\bar{x}) = \mu$, $V(\bar{x}) = \sigma^2/n$. This satisfies the above conditions and is thus a consistent estimator. As n increases the distribution of $\bar{x}$, which is also normal, gets narrower and narrower so the observed $\bar{x}$ tends, in a probability sense, to be nearer and nearer to μ.

Chapter 15

Eclectic approaches to estimation*

15.1 Introduction

The aim of this chapter is to indicate briefly how we might bring together the prior information and the data to obtain estimates. Some methods of bringing together the ideas of the two previous chapters will be left until the study of iteration in Chapter 21 since they are often applied as a second stage after initial estimates have been developed. In this chapter we shall consider approaches which are based on an inherent joining of the conceptual and the empirical. The usual way of doing this is by the use of Bayesian methods, and we devote most of this chapter to examining these methods.

15.2 Bayesian methods

In Chapter 12 we introduced Bayes formula which enables the inversion of conditional probability formulae:

$$Pr(M_i|D) = Pr(D|M_i)\,Pr(M_i)/Pr(D),$$

where $Pr(M_i)$ and $Pr(M_i|D)$ are the probabilities of a model M_i being true unconditionally and given available data D respectively. $Pr(D)$ and $Pr(D|M_i)$ are the probabilities of the data unconditionally and given the model M_i respectively. The model M_i is one of a specified group of models $M_1, M_2, \ldots, M_N$. In the context of this chapter we assume that we know the general form of the model, $f(x; \theta)$. The choice is then between the various possible values for θ. We thus have the following change of notation.

(a) $Pr(D|M_i) \rightarrow Pr(x_1, x_2, \ldots, x_n|\theta)$

for a discrete distribution or

$$Pr(D|M_i) \rightarrow f(x_1, x_2, \ldots, x_n|\theta)$$

for a continuous distribution. For a set of independent observations we can write this as

$$f(x_1;\theta)\,f(x_2;\theta)\,\ldots\,f(x_n;\theta).$$

Note that if we require this expression in terms of θ this is in fact the likelihood, $\mathscr{L}(\theta)$.

(b) $Pr(M_i) \to h(\theta)$. This is the prior distribution for θ.
(c) $Pr(D) \to f(x_1, x_2, \ldots x_n)$, the unconditional, marginal, distribution of the data.
(d) $Pr(M_i|D) \to h(\theta|x_1, x_2, \ldots, x_n)$, the posterior distribution of θ given the data $x_1, x_2, \ldots, x_n$.

Rewriting Bayes formula for this new situation where we have prior knowledge of the parameter in the form of a prior distribution and a set of data from a known distribution, $f(x; \theta)$, gives

$$h(\theta|x_1, x_2, \ldots, x_n) = \{f(x_1;\theta)\,f(x_2;\theta)\,\ldots\,f(x_n;\theta)\}\,h(\theta)/\,f(x_1, x_2, \ldots x_n).$$

Writing $\{\ \ \}$ as the likelihood and $1/f(x_1, x_2, \ldots, x_n)$ as just a multiplying constant C gives

$$h(\theta|x_1, x_2, \ldots, x_n) = C\mathscr{L}(\theta)\,h(\theta),$$

in words

Posterior Distribution $=$ Constant $\times$ Likelihood $\times$ Prior Distribution

To illustrate this result we look at a number of examples.

Example 1
Consider the Poisson Process as a model for faults in a production process. The fault rate is the parameter of interest and is denoted by v faults per unit time. The process is observed for a time t and r faults observed. Consider again the elements in the Bayes formula:

(a) The distribution—the likelihood
In this situation the number of faults in time t has expectation vt and is distributed as a Poisson so

$$Pr(r|v) = e^{-vt}\,\frac{(vt)^r}{r!} = \mathscr{L}(v).$$

(b) The Prior Distribution
We need to model the prior distribution of v by a distribution over $(0, \infty)$. One such distribution is the Gamma distribution which in a rather general form can be written as

$$h(v) = T(Tv)^{T\alpha-1}\,e^{-Tv}/\Gamma(\alpha T), \qquad 0 < v < \infty$$

$$= \text{Const. } v^{T\alpha-1}\,e^{-Tv}. \qquad 0 < v < \infty$$

The quantity $\Gamma(\alpha T)$ is here a constant called the Gamma function, that

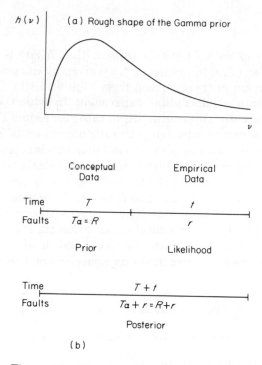

Figure 15.1 The Gamma prior and its use

we need not worry about in the actual use we make of this formula. This distribution is sketched in Figure 15.1(a). The mean of this distribution, the expectation of v, is

$$E(v) = \alpha.$$

So the expected number of events in a time T is $R = T\alpha$. The value of choosing this distribution will become apparent.

(c) The Posterior Distribution

Bayes formula now becomes

$$h(v|r) = C . e^{vt} \frac{(vt)^r}{r!} . T(Tv)^{T\alpha-1} e^{-Tv}/\Gamma(vT),$$

which, on rearranging and putting all the constant terms together as C', is

$$h(v|r) = C' v^{r+T\alpha-1} . e^{-(T+t)v}$$

If we write

$$T^* = T + t$$

$$\alpha^* = \frac{r + T\alpha}{T + t} = \frac{R^*}{T^*},$$

where $R^* = R + r$.

This becomes

$$h(v|r) = C' \, v^{T^*a^*-1} \, e^{-T^*v}.$$

Comparing $h(v|r)$ with $h(v)$ it is evident that $h(v|r)$ is also a Gamma distribution with (T, α) becoming (T^*, α^*) given by the previous relations. The interpretation of these is seen from Figure 15.1(b). Our prior distribution represents a conceptual experiment in which we imagine the process running with faults occurring at rate α for a time T. Our choice of α reflects our view on the expected fault rate, our choice of T our confidence in this assessment. If we are not very confident this is represented by a very short conceptual experiment with T quite small relative to t. If we are very confident we might allow T to be bigger than t. The use of Bayes formula effectively combines this *Conceptual Data* (R, T) with the actual *Empirical Data* (r, t) to give a combined set. This posterior information corresponds to observing $R + r$ faults in a total observation time $T + t$.

Prior distributions that, with the likelihood, lead to posterior distributions of the same form are called *Conjugate Priors*. Table 15.1 gives a list

Table 15.1 The interpretation and use of conjugate priors

---

1. Model: random events over time t, mean rate v
   $R_t \cap$ Poisson (vt)
   Prior: $h(v) \cap$ Gamma (T, α)
   Conceptual Data: Time T showing R events, $(R = T\alpha)$
   Empirical Data: Time t showing r events
   Total Data: Time T^* showing R^* events
   $$T^* = T + t, \; R^* = R + r, \; R^* = T^*\alpha^*$$
   Posterior: $h(v|r) \cap$ Gamma (T^*, α^*)
2. Model: Binomial situation Bin (n, θ)
   Prior: $h(\theta) \cap$ Beta (α, β)
   Conceptual Data: $\alpha + \beta$ trials with α successes, β failures
   Empirical Data: n trials with r successes, $n - r$ failures
   Total Data: $\alpha^* + \beta^*$ trials with α^* successes, β^* failures

   $$\alpha^* = \alpha + r, \; \beta^* = \beta + n - r.$$

   Posterior: $h(\theta|r) \cap$ Beta (α^*, β^*)
3. Model: Normal, known variance $N(\mu, \sigma^2)$
   Prior: $h(\mu) \cap N\left(\mu_0, \dfrac{\sigma^2}{m}\right)$
   Conceptual Data: m observations with mean μ_0
   Empirical Data: n observations with mean $\bar{x}$
   Total Data: m^* observations with mean μ_1^*

   $$\mu_1^* = \frac{m\mu_0 + n\bar{x}}{m + n}, \; m^* = m + n$$

   Posterior: $h(\mu|x_1, \ldots, x_n) \cap N\left(\mu^*, \dfrac{\sigma^2}{m^*}\right)$

---

of some of these together with the interpretation of the Conceptual Data and Empirical Data.

Thus far we have not referred to estimators. In estimation we seek not a posterior distribution of θ but a single value, or some form of confidence interval. However, such values come naturally from the posterior. To obtain these estimators we use criteria related to the posterior distribution, often called *Loss Functions*. For example:

$$\text{The Mean Square Error MSE} = E\left[(\theta - \hat{\theta})^2\right]$$

$$\text{The Mean Absolute Error MAE} = E\left[|\,|\theta - \hat{\theta}1|\,\right].$$

These differ from their previous use since $E[\ ]$ is obtained from the posterior distribution. If for example we require an estimator with minimum mean square error then such an estimator is provided by the expectation of the posterior distribution. Thus in the previous example α^* is the optimum estimator of v in the MSE sense. It can similarly be shown that if the criterion of a good estimator is the minimum mean absolute error then the median of the posterior distribution is the best estimator.

If an interval is required then this can be obtained directly from the posterior distribution. For example the values (θ_L, θ_U) that enclose the central 95% of the distribution provide a 95% probability interval for θ.

Example 2

Consider the situation where a series of observations are taken with an instrument with known $\sigma = 1.60$ units. Twenty observations are taken and the sample mean is

$$\bar{x} = 293.00 \text{ which as variance } \frac{\sigma^2}{20} = 0.128.$$

Consider two possible priors both $N\left(\mu_0, \frac{\sigma^2}{m}\right)$. Our previous experience suggests $\mu_0 = 293.40$. For prior (a) we assume a considerable confidence in this value so our conceptual experiment is taken as corresponding to 20 observations. It follows from Table 15.1 that the posterior is

$$N\left(\frac{20 \times 293.0 + 20 \times 293.4}{20 + 20}, \frac{1.6^2}{20 + 20}\right)$$

$$= N(293.20, (0.250)^2).$$

For prior (b) we assume much less confidence and set $m = 1$, we then have the posterior of

$$N\left(\frac{20 \times 293.0 + 1 \times 293.4}{20 + 1}, \frac{1.6^2}{21}\right) = N(293.02, (0.35)^2).$$

So in this case the sample data dominates the 'prior data'.

Remembering that 95% of a normal distribution lies within 1.96 standard deviations of the mean it is seen that the 'Bayesian' 95% probability limits for μ are

$$\text{(a)} \quad (292.71, \ 293.69),$$

and

$$\text{(b)} \quad (293.33, \ 293.71).$$

The limits (b) are very similar to the 95% confidence limits since almost all the evidence comes from the empirical data, putting $m = 0$ would give identical limits. Note, however, that (b) is a statement about probabilities of μ, whereas we were careful not to use such terminology in defining confidence limits. In the Bayesian approach parameters *are* treated as random variables.

The previous two examples have illustrated the intuitive value of conjugate priors. In section 8.7 we introduced a 'vague' prior that was uniform. Suppose we have a uniform prior $U(\alpha, \beta)$ then the posterior

$$h(\theta|D) \propto \mathscr{L}(\theta) \cdot \frac{1}{\beta - \alpha}, \quad \begin{array}{l} \alpha < \theta < \beta \\ \text{otherwise} \end{array}$$

The prior represents the prior information that no values of θ in (α, β) are more probable than any other. If α and β are so wide that outside that range $\mathscr{L}(\theta)$ is negligibly small we can write as a good approximation

$$h(\theta|D) \propto \mathscr{L}(\theta), \quad \text{all } \theta.$$

Thus the $\mathscr{L}(\theta)$ acts as the posterior distribution, and the prior tells nothing of relevance about θ. Priors that convey no information are termed non-informative priors. Such a prior need not be a uniform distribution it may be sufficient for it to be relatively flat in the range where $\mathscr{L}(\theta)$ is non-negligible.

To obtain a proper posterior p.d.f. for θ the likelihood must be standardized to give unit area under $\mathscr{L}(\theta)$.

Example 3
Returning again to the regression model

$$y_i = \theta x_i + e_i,$$

and assuming now that $e \cap N(0, \sigma^2)$, σ^2 known, the likelihood is obtained from the argument used in subsection 13.4.1 as

$$\mathscr{L}(\theta) \propto \exp - \frac{1}{2\sigma^2} \sum e_i^2$$

$$\propto \exp - \frac{1}{2\sigma^2} \sum (y_i - \theta x_t)^2.$$

If $\hat{\theta}$ is the standard least squares/maximum likelihood estimator of θ we can write

$$
\begin{aligned}
\sum (y_i - \theta x_i)^2 &= \sum [(y_i - \hat{\theta} x_i) - (\theta - \hat{\theta}) x_i]^2 \\
&= \sum [\hat{e}_i - (\theta - \hat{\theta}) x_i]^2 \\
&= \sum \hat{e}_i^2 + (\theta - \hat{\theta})^2 \sum x_i^2.
\end{aligned}
$$

The cross product term involves $\sum \hat{e}_i x_i$ which is zero by the normal equation. Thus

$$
\mathscr{L}(\theta) = \exp - \left\{ \frac{\sum x^2}{2\sigma^2} (\theta - \hat{\theta})^2 \right\}
$$

and hence if θ has a non-informative uniform prior the posterior will be proportional to this expression. From the form of this it is seen that this posterior distribution is

$$
N \left(\hat{\theta}, \frac{\sigma^2}{\sum x_i^2} \right),
$$

i.e. a normal distribution centred on the Least Squares estimator. It follows that the Least Squares estimator is also the minimum mean square error, $E_\theta [(\theta - \hat{\theta})^2]$, estimator, and that the Bayesian 95% interval is $\left(\hat{\theta} \pm 1.96 \frac{\sigma}{\sqrt{\sum x^2}} \right)$.

15.3 Fitting models of complex systems

One area where the bringing together of conceptual and empirical approaches to estimation has been extensively used is the study of complex systems, such as in the use of System Dynamic Models. Fitting such models can be approached in three stages:

(a) a preliminary fit of the modules of the model,
(b) an adjustment of fitted parameters to get the overall model behaviour realistic, and
(c) an investigation of the sensitivity of the model to variability in the parameters, which corresponds to the study of efficiency in the previous chapter.

Considering these briefly in turn:

(a) The modular nature of systems often enables the separate modules to be modelled and fitted independently. This may be done from available data or from trials and experiments set up for the purpose. In other cases the conceptual approach to constructing the models tends to lead to parameters that have a clear interpretation, in terms of the prototype, and that can be 'guesstimated' by the expert in the appropriate field of study. The procedure of estimating all the parameters at once from all the data, that has been used in regression and most other models, is thus rarely carried out for the more complex models; partly because of their very complexity, in particular their non-linear structures, partly because it ignores the substantial and varied prior information that must exist before the models can be constructed, and partly due to lack of complete sets of adequate and comparable data for all the model variables.

(b) Having obtained a first fit of the model the behaviour of the model can be simulated on a computer. The inputs to drive the model may be (i) historical sets of data for exogenous variables, or (ii) mathematical test functions such as step changes, linear trends or sine curve oscillations (iii) random sequences or (iv) appropriate combinations of (i), (ii) and (iii). The model behaviour under such simulations is examined. Though the short-term predictive quality may be of some interest it must be remembered that these models are used mainly as tools for investigating system behaviour. Thus criteria like residual sums of squares are inappropriate. Instead of asking how close are the model-generated numbers to observed numbers, the questions become:

   (i) Are any oscillations shown by historical data also shown, in general shape and period, by data simulated from the model?

  (ii) Are trends, turning points, growth, decay patterns similar?

 (iii) Do changes in any decision variables have the same type of effect as observed in practice?

 (iv) Are the general forms of relationship between the variables similar in simulation and prototype?

  (v) If the model is started with a state corresponding to that of some state of the prototype, do they behave similarly thereafter, both in qualitative and in approximately quantitative terms.

These types of question lead to a much broader approach than that produced by defining a single criterion of fit. Consequently, it is found that a range of parameter values can lead to equally acceptable simulations, so behaviour is not too sensitive to the estimates used. Nonetheless, these studies enable the initial estimates to be adjusted to achieve better model behaviour.

(c) With simple models we can often quote standard errors of estimated parameters, so the user has some feel for the precision of his fitted model. With complex models, particularly system dynamics models, it is very difficult to estimate the parameters themselves and little attempt is made to move to the additionally difficult level of finding standard errors. Indeed the fact that such models tend to lead to estimates with strong intercorrelations, makes the interpretation of standard errors difficult. A consequence of these difficulties is that complex models are often fitted and presented without any measures of uncertainty. They often therefore give a false sense of reliability. Many complex situations are characterized by a shortage of data and there is a strong suspicion that quoted estimates of parameters are very imprecise. An approach that seeks to explicitly acknowledge this problem is to use simulated data to investigate sensible ranges of parameter values, rather than to seek point estimates. For example, a study by Fedra, Van Straten and Beck of a lake model uses available data and prior knowledge to set reasonable ranges of values for: (i) model parameters, (ii) input variables to the lake system, (iii) responses of the lake system to the inputs.

Values for both (i) and (ii) were simulated using the uniform prior distribution and the model used to find the responses. These responses were then examined to see whether they lay in the range thought reasonable. Figure 15.2 indicates the nature of the technique. In their study the authors had 10 parameters, 12 input variables, X, and 7 response variables, Y, describing the systems behaviour. Thus the two-dimensional illustration of Figure 15.2 represents the typical multidi-

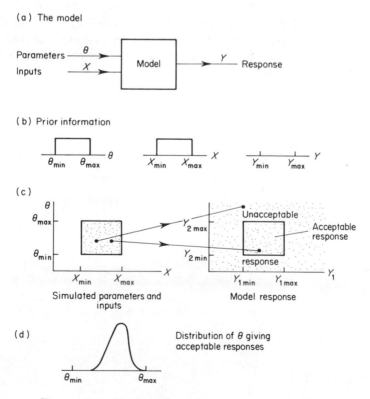

Figure 15.2 A simulation study of parameter values

mensional analysis required. In the study 10,000 simulations were used and only 300 led to plausible responses. An analysis of the model parameters used in these 300 acceptable simulations then gave a clear view, of the ranges of actual parameter values that were consistent with the observations used in setting up the simulation. A by-product of this investigation was the discovery of very high correlation between some parameter values in the 300—which implies that the model contains more parameters than necessary. Thus though this example was introduced as a way of investigating the quality of estimates in a systems model it also raises questions about the validity of the model used. Such questions are the subject matter of the next part of our study of statistical modelling.

Before leaving this topic it is necessary to underline that we have been discussing large complex models and we have emphasized the problems associated with their complexity. Largeness as such, in terms of numbers of variables, does not pose such problems if the form of the model is simple. There exist linear multiple regression models, simultaneous equation models, multivariate ARMA models, etc., involving many variables. Their structure, however, is essentially the same as that described in Part II of this book and they can be fitted essentially by the methods of estimation discussed in Chapter 13. Additional problems relate only to the handling of large quantities of data and the design of efficient computer algorithms for numerically finding maximum likelihoods when there are many parameters in the likelihood function.

Part V

Validation

Chapter 16

The elements of validation

16.1 The nature of validation

For a young child two small pieces of wood joined in a cross form a perfectly adequate model of an aeroplane. As the child gets older the discrepancy between model and reality becomes clearer and model aircraft that fly and scale models are used to identify with real aircraft. The process of comparison of the model with the observed world is called *Validation*. What is valid at one stage of a study may not be seen as valid at a more developed stage. A model may be valid for one purpose but not for another. The object of validation is thus to examine whether the model is a good (not true) description of a prototype in terms of its behaviour and of the application proposed.

As so often in our study of statistical modelling, we are not dealing with a topic that can be treated on the basis of a few theorems and optimum results, at least not yet. Rather we are in a relatively vague and unexplored area with only a few signposts to guide us. The aim of this part of our study is to discuss and illustrate some ideas and methods of value in validation. As usual it is helpful to distinguish between the conceptual and empirical approaches. The essence of validation is the comparison of the model and its behaviour with the prototype. Information about the prototype consists of a knowledge base and a data base. Comparison of the model with the knowledge base, with our prior information, will be called *Conceptual Validation*, comparison with the data base *Empirical Validation*. The attempt to combine both sources of information we will call *Eclectic Validation*. In taking the conceptual approach we must ask whether the model behaves the way that we would conceive it as behaving, does it match our prior knowledge of the way the prototype behaves. If we take an empirical approach we must compare the behaviour of the model with actual observations on the prototype. To carry out an effective comparison we need to bring together a number of elements:

(a) a checklist giving the knowledge base for the prototype and if we are doing empirical validation we need the data base appropriate;

(b) information on the proposed use of the model;
(c) a checklist of model properties.

The nature of the data from the prototype that will be used for empirical and eclectic validation needs careful thought. Such data can be of three types:

Fitting Data—this is data that has already been used in the fitting of the model and possibly also in the identification process.
Validation Data—this is data collected especially for the validation process.
Working Data—this is data arising in the day to day use of a model.

How validation is carried out will have significant differences for the three types of data. Intuitively we are more likely to appear successful in validating a model with fitting data than with validation data. The fitted model will fit more snugly to the fitting data than to any validating data, so we might incorrectly accept a poor model or begin to expect too much of a good model. The thoughtless use of fitting data can thus be misleading. Validation data provides a much more stringent basis for comparison of model and prototype. Working data presents particular problems since usually it will be gathered in a continuous fashion and the validation will be carried out as a routine procedure.

Producing a checklist of model properties requires a systematic analysis of the model. For the purposes of discussion we will subdivide this analysis into four categories:

Form analysis, which is the study of the structure and implicit assumptions of the model.
Behavioural analysis, which is the study of the expected, probabilistic and asymptotic behaviour of the model.
Sensitivity analysis, which is the study of the influences of errors or changes in the model components.
Misspecification analysis, which is the study of the effects of using an incorrect model in place of an assumed correct model.

The objective of this chapter is to study and illustrate these various elements in the validation process. In the next chapter we shall discuss how to carry out the comparisons of the validation process.

16.2 Prototype analysis

A common cause of error in modelling is the use of a model devised for one prototype in another situation that in the event turns out to have some important differences. Thus, even if a thorough analysis of the prototype was carried out at the conceptual identification stage, an updated checklist of prototype properties should be created at the validation stage. Obviously, if a model is to be used that was developed by others for a different prototype then a checklist for the new prototype will have to be produced. The contents of such checklists have already been discussed in Chapter 8.

Closely linked with consideration of the prototype properties is the examination of the environment of the prototype and how the prototype was observed. The classic distinction here is between the prototype observed and measured as part of a carefully designed experiment, in a controlled environment, and apparently the same prototype observed and measured in its natural, uncontrolled, environment.

Example 1
Consider a situation modelled by

$$y = \alpha + \beta x + e,$$

in which other variables u, v, w exist but are not included in the model. In an experimental study we ask how x influences y. We control u, by keeping it fixed at u_0, and the environment is controlled to keep out effects from v and w. The fitted model describes the experimental prototype and environment. In an observational study, say a survey, the variable u fluctuates and has a strong relation to both x and y. Hence the regressor x acts on its own behalf and also as a proxy for u. Variables v and w influence y independently of x and their effect is included in the error term e. Thus the model now is seen as appropriate for a very different prototype and environment. Though the experimentally derived fitted model may give a better picture of the actual relation between x and y, it may well be that it gives a misleading picture, if u is not u_0 and if the fluctuations of v and w do influence y. In this latter case it would be better for the coefficient β to allow for the proxy role of u and for the estimated error variance to include the influence of v and w. So, though the model is the same, and the variables are the same, the prototypes and environments being modelled are fundamentally different.

16.3 Application analysis

In practice the validation stage will come before the application stage, though an experimental or trial application may well provide information and data of relevance to validation. It is, however, important, in considering the approach to validation, to give thought to the nature of the application. Certainly a failure to do this can lead to a misleading validation. The reason for this is that a validation based only on consideration of the prototype and the model, will lead to an approach that emphasizes detailed analogy between the two. The tendency will be to favour complex models, since these give the maximum correspondence between the model and the complexities of real life prototypes. The evidence of experience in some areas of application, however, is that once one starts to apply the model then complex models do not necessarily perform better than simple models. It is the nature and requirements of this performance that need examining. The classic example is the application of models to forecasting. In this case the success of the model depends solely on the closeness of the forecasts to the observed values. There is thus no requirement at the validation stage for the form of the model to resemble the form of the prototype. The focus of the validation in this case needs to be on the evidence of predictive ability for the proposed model. The nature of the application thus has a role in determining the relative importance of the various components in the prototype and model checklists. In the case of empirical validation the application will often suggest the numerical criteria appropriate for the validation.

Example 1
Suppose that the output of a chemical process, y, is modelled by a regression model.
$$E(y) = \alpha - 2\gamma\theta x + \gamma x^2,$$

The variable x is the process temperature, the parameters are expressed in a convenient fashion and α and γ are assumed known from chemical theory. The maximum of $E(y)$ is obtained by setting $x = \theta$ which gives the maximum output of $\alpha - \gamma\theta^2$. Given a set of data we will estimate θ by an estimator $\hat{\theta}$. The natural measure of the quality of $\hat{\theta}$ is the loss, $l(\hat{\theta})$, in output produced by using $x = \hat{\theta}$ instead of $x = \theta$. This is
$$l(\hat{\theta}) = (\alpha - \gamma\theta^2) - (\alpha - 2\gamma\theta\hat{\theta} + \gamma\hat{\theta}^2)$$
$$= \gamma(\theta - \hat{\theta})^2.$$

Thus the criterion, the loss function, for judging $\hat{\theta}$ is the square of its error.

16.4 Model form analysis

The objective of *Form Analysis* is to make explicit in a checklist the detailed nature of the model being validated to facilitate the comparison with the analogous prototype checklist. It will often happen that in carrying out the modelling process we are unconsciously led to make assumptions that were not there at the beginning. Such assumptions are made explicit in form analysis.

Example 1
A modelling exercise with dependent variable y and several regressors M, P, T, W leads to the model
$$y = \beta_0 + \beta_1 M + \beta_2 P + \beta_3 MP + e.$$

In checklist form this model implies:

  (i) Variables M and P separately have a linear influence on y.
 (ii) Variables T and W are not necessary to describe y's behaviour.
(iii) There is an interaction, MP, between M and P in their influence on y.
(iv) The random error in the situation can be treated as additive.
 (v) The model has a constant term.

16.5 Model behaviour analysis

There are as many ways of analysing and describing model behaviour as there are models, but some of these constantly re-occur. These include analysis of:

(a) 'normal' behaviour based on expectations or averages,
(b) 'probable' behaviour or shape based on probability distributions or intervals,
(c) asymptotic behaviour as time increases or generations pass
(d) stable, unstable or oscillatory behaviour, which often relates to (c).

The three main methods of investigating model behaviour are:

(a) by mathematical analysis obtaining exact formulae describing the behaviour,
(b) by solving more complex model equations using numerical approximations,
(c) by simulating normal model behaviour, probably on a computer.

In the following subsections we discuss and illustrate some methods of model behaviour analysis.

16.5.1 Expected behaviour

The statistical idea of the expectation provides a natural tool for examining normal, or expected, behaviour. For distributional models $E(X)$, $V(X)$ and the higher moments provide natural statements about the model. For linear models, with $E(e) = 0$, the deterministic part of the model provides the expected behaviour. For progression and non-linear models an analysis of expected behaviour requires more careful investigation. The following examples illustrate some of the techniques and ideas.

Example 1
Consider the autoregressive model AR(1) given as

$$x_t = \phi x_{t-1} + e_t, \qquad E(e_t) = 0, \qquad V(e_t) = \sigma^2,$$

with ϕ numerically specified.

If x_{t-1} is known then ϕx_{t-1} is a constant so the conditional expected value of x_t given x_{t-1} is

$$E(x_t|x_{t-1}) = \phi x_{t-1} + E(e_t) = \phi x_{t-1}.$$

If x_t is predicted using ϕx_{t-1} the error will be

$$x_t - \phi x_{t-1} = e_t$$

so the predictive errors will have the basic properties of e_t itself and can thus be used for a graphical validation, using the techniques developed for residuals.

Using γ_h to denote the autocovariance of lag h and using the model twice gives

$$x_t x_{t-1} = \phi x_{t-1}^2 + e_t x_{t-1}.$$

Taking expectations and using the independence of e_t and x_{t-1}, so that $E(e_t x_{t-1}) = E(e_t) E(x_{t-1}) = 0$, we have

$$\gamma_1 = \phi \sigma_x^2,$$

assuming the variance of x remains constant at σ_x^2. Similarly

$$x_t x_{t-h} = \phi x_{t-1} x_{t-h} + e_t x_{t-h}$$

so

$$\gamma_h = \phi \gamma_{h-1}.$$

Hence

$$\gamma_h = \phi^h \sigma_x^2$$

and the autocorrelation ρ_h for observations h apart is

$$\rho_h = \gamma_h / \sigma_x^2 = \phi^h,$$

which for $|\phi| < 1$ is a decreasing function of h which oscillates if ϕ is negative. Squaring the equation of the model gives

$$x_t^2 = \phi^2 x_{t-1}^2 + 2\phi x_{t-1} e_t + e_t^2,$$

hence

$$\sigma_x^2 = \phi^2 \sigma_x^2 + \sigma_e^2,$$

and

$$\sigma_x^2 = \sigma_e^2 / (1 - \phi^2).$$

Thus for small ϕ the variability of x is little larger than the variability of e.

Example 2

In a study of the density of a larva in successive generations the following linearized equation arose

$$\ln N_t = \alpha + \beta \ln N_{t-1} + e_t$$

when N_t is the density in generation t of larvae per ten square feet of foliage. Writing in the non-linear form gives

$$N_t = \alpha^* N_{t-1}^\beta u_t$$

where $\alpha = \ln \alpha^*$ and $e = \ln u$. Thus if e has a $N(0, \sigma^2)$ distribution u has a lognormal distribution with expectation $e^{\frac{1}{2}\sigma^2}$. Assuming u_t is independent of previous N_{t-1} leads to

$$E(N_t | N_{t-1}) = \alpha^* N_{t-1}^\beta e^{\frac{1}{2}\sigma^2}.$$

Thus the expected density depends on the previous density but is larger than indicated by the deterministic part of the model.

16.5.2 Asymptotic–steady state behaviour

Many models show characteristic long-run limiting behaviour, such as asymptotic growth, or the steady state oscillations, shown by many physical systems and system dynamics models. Again we illustrate by example.

Example 3
Consider again the model for the growth of a population of larvae.

$$\ln N_t = \alpha + \beta \ln N_{t-1} + e_t.$$

The random term in the model implies that even if a steady state occurs there will be local random variation about it. We can represent this by defining a steady state N such that

$$\ln N_t = \ln N + z_t,$$

where z_t is the local random variation. Substituting in the model

$$\ln N + z_t = \alpha + \beta(\ln N + z_{t-1}) + e_t.$$

Separating the deterministic and stochastic components to give equality for all t we will have the steady state N satisfying

$$\ln N = \alpha + \beta \ln N,$$

so

$$N = e^{\alpha/(1-\beta)}.$$

The stochastic element will then be given by

$$z_t = \beta z_{t-1} + e_t,$$

which is an Autoregressive, AR(1), model.

Example 4 (A deterministic progression model)
In examining model behaviour it is often useful to redefine the model parameters.

In Chapter 9, Table 9.1, we derived a model for an electrical circuit in the form

$$LC\frac{d^2V}{dt^2} + RC\frac{dV}{dt} + V = V_{input}.$$

If we now write $LC = 1/w^2$ and $RC = 2\xi/w$ the model becomes

$$\frac{d^2V}{dt^2} + 2\xi w\frac{dV}{dt} + w^2V = w^2V_{input}$$

A standard result in calculus is that the general solution form of this model setting $V_{input} = 0$ after an initial disturbance is

$$V = Ae^{-\xi wt}[e^{w\sqrt{(\xi^2-1)}t} + Be^{-w\sqrt{(\xi^2-1)}t}] \qquad \text{if } \xi > 1$$
$$V = Ae^{-\xi wt}(1 + Bt) \qquad \text{if } \xi = 1$$
$$V = Ae^{-\xi wt}\sin(w\sqrt{(1-\xi^2)}t - \phi) \qquad \text{if } 0 < \xi < 1$$
$$V = A\sin(wt - \phi) \qquad \text{if } \xi = 0.$$

Thus the behaviour of the model is

(i)   a decay to zero if $\xi \geqslant 1$,
(ii) oscillatory if $0 < \xi < 1$ with the oscillation dying away to zero,
(iii) if $\xi = 0$ then the behaviour is a simple oscillation.

Thus the parameter ξ, called the *Damping Factor*, determines the form of the behaviour of the model. The steady state in particular is a 'dead state' $V = 0$ if $\xi > 0$ but an oscillation of period $2\pi/w$ if $\xi = 0$, thus w determines the period of the undamped oscillation. The old parameters L C and R described the component properties but the redefined parameters ξ and w describe the system properties.

16.5.3 Probabilistic behaviour

The simplest way of describing the probabilistic nature of a model is by use of simple probability limits

Example 5
The load-bearing strength of some reinforced concrete pillars is modelled by a normal distribution of mean, μ, 400 tons and standard deviation, σ, 40 tons. Using the symmetry of the normal distribution, $N(0, 1)$, and standard Tables we see that

$\mu \pm 0.6745\sigma = (373,427)$ will contain half the population
$\mu \pm 1.6449\sigma = (334,466)$ will contain 90% of the population
$\mu \pm 1.9600\sigma = (322,478)$ will contain 95% of the population
$\mu \pm 2.5758\sigma = (297,503)$ will contain 99% of the population

Such a set of intervals gives a reasonable feel for the probability behaviour of the model.

The above example assumes that we have a perfectly fitted model and that we are simply looking at the distribution. In the practice of validation we will often have models based on limited data and our concern will be to compare what happens, with what the fitted model predicts will happen. Let us extend the above example to this situation.

Example 6

Let us suppose that a sample of five concrete pillars are loaded to breaking point and the sample mean and standard deviation obtained were 400 tons and 40 tons respectively. The analysis is clearer if we express the model as

$$x = \mu + e,$$

where x is the load at failure, μ is the mean strength and e the deviation from that load. By the model $e \cap N(0, 40^2)$. For the fitted model we have

$$x = \hat{\mu} + \hat{e}.$$

As $\hat{\mu}$ is the average of five observations

$$\hat{\mu} \cap N\left(\mu, \frac{\sigma^2}{5}\right).$$

The predicted value of a new, validation, observation is

$$\hat{x} = \hat{\mu}.$$

The actual value is

$$x = \mu + e.$$

The error in the prediction is thus

$$e_p = x - \hat{x} = (\mu - \hat{\mu}) + e.$$

The expected value of this error is zero and its variance is given by

$$\begin{aligned} V(e_p) &= V(\mu - \hat{\mu}) + V(e) \\ &= V(\hat{\mu}) + V(e) \\ &= \frac{\sigma^2}{5} + \sigma^2 \\ &= 1.20\sigma^2 \end{aligned}$$

so

$$\sigma(e_p) = 1.095 \times 40 = 43.8.$$

Thus the standard deviation of the prediction error is about 10% higher than that indicated at first sight by the fitted model. If we wish to put normal probability limits round the predicted value these will be wider by 10%, because of the use of the estimated $\hat{\mu}$ in place of the true μ. However, we are not only estimating μ but also σ. The multipliers used to obtain the interval were based on

$$\frac{\bar{x} - \mu}{\sigma/\sqrt{n}}$$

having a N(0, 1) distribution. Now we replace σ by s and use

$$\frac{\bar{x} - \mu}{s/\sqrt{n}}$$

having a t-distribution with four degrees of freedom (d.f.).

Tables of the t-distribution show that the 90 % limits of 1.64 for the N(0, 1) distribution become 2.13 for the t-distribution with 4 d.f. This is a major change due to the lack of sufficient data to obtain a reliable estimate of σ. Combining the above two results leads to a 90 % interval of

$$400 \perp 2.13 \times 43.8 = (307, 493)$$

as the predicted interval for a further observation. Compare this with the 90 % interval of (334, 466) for the distribution N(400, 40^2). This illustrates the general result that low precision in estimation can have substantial effects on the precision with which the fitted model can be used to predict further observations and hence on comparisons made with validation data. Obviously as the amount of data increases both estimators get closer to the true values and the interval approaches the standard normal interval.

16.6 Model sensitivity analysis

The objective of sensitivity analysis is to study how changes in the prototype or the environment would effect the behaviour of the model, for example through changes in variables, parameters, or model structure.

Example 1

In a regression model changes or errors in parameter values, β, will have a direct effect on the dependent variable, y. The rate of change of y with β is called the *Sensitivity Coefficient* c_β. Consider the model

$$y = \alpha + \beta x + e.$$

The two sensitivity coefficients are found by differentiating with respect to α and β, which are denoted by

$$c_\alpha = \frac{\partial y}{\partial \alpha} = 1$$

and

$$c_\beta = \frac{\partial y}{\partial \beta} = x.$$

It is helpful to express these in the form of a *Sensitivity* Elasticity which gives the percentage rate of change in the expected y per unit percentage change in the parameter. These are,

$$l_\alpha = c_\alpha \frac{\alpha}{E(y)} = \frac{\alpha}{\alpha + \beta x},$$

$$l_\beta = c_\beta \frac{\beta}{E(y)} = \frac{\beta x}{\alpha + \beta x}.$$

Example 2
Returning to the growth model for larvae the steady state population is

$$N = e^{\alpha/(1-\beta)}.$$

The sensitivity coefficients are thus

$$c_\alpha = \frac{\partial N}{\partial \alpha} = \frac{1}{1-\beta} N$$

$$c_\beta = \frac{\partial N}{\partial \beta} = \frac{-\alpha}{(1-\beta)^2} N.$$

Hence the sensitivity elasticities are

$$l_\alpha = c_\alpha \cdot \frac{\alpha}{N} = \alpha/(1-\beta)$$

$$l_\beta = c_\beta \cdot \frac{\beta}{N} = -\alpha\beta/(1-\beta)^2.$$

The relation between l_α and l_β is

$$l_\beta = -\{\beta/(1-\beta)\}\, l_\alpha.$$

Thus for β close to one the effect of error in β will be much larger than that for error in α.

A further aspect of sensitivity analysis is the examination of how sensitive the model behaviour is to (a) changes in the input or exogenous variables of the prototype and (b) those aspects of the environment, those other variables, that were left out of the final model as being of little importance or unmeasurable.

Example 3
In the development of inertial guidance systems, models of the prototype system will have as model variables the various forces and accelerations acting on and in the system. The environment is made up of all the input vibrations occurring during flight, small forces neglected in model

construction, local variations in the earth's gravitational field, etc. Over repeated development of such systems, models of this external environment and of the input variables have been constructed. Such models are called *Reference Models*. Using a reference model it is possible to carry out flight simulations, in which the reference model creates a simulated environment and flight path and the system model does its job of estimating the current flight position. An examination of the error between simulated and estimated positions gives a means of studying model quality. It is then possible to examine the effect on error of different factors in the reference model. Such an analysis of sensitivity to different factors leads to an *Error Budget* that breaks the total error in some feature, e.g. north position error, into percentage contributions from various contributory factors, such as gyro drift rate, gravity anomalies, and noise and bias in accelerometers. The reference by Maybeck discusses such techniques.

16.7 Model misspecification analysis

As a final consideration we turn to the implications of model misspecification. In general this is a topic about which very little is known. There has been some study in the context of regression models, distributions, and ARMA models, but even here it has been very much a matter of investigating illustrative examples. The best advice to the modeller is to carry out an investigation of his own models. The simplest case to deal with is that of nested models, for here the form of misspecification is either an overspecification, with too many terms included, or an underspecification with too few terms included.

Example 1
Consider two simple regression models

$$y = \theta x + e \tag{A}$$

$$y = \beta_1 x + \beta_2 z + u \tag{B}$$

First consider the case of underspecification in which we use (A) when (B) is the appropriate model. If we apply Least Squares to both models we obtain the Normal equations

(A) $$\sum xy = \hat{\theta} \sum x^2$$

(B) $$\left\{ \begin{array}{l} \sum xy = \hat{\beta}_1 \sum x^2 + \hat{\beta}_2 \sum xz \\ \sum zy = \hat{\beta}_1 \sum xz + \hat{\beta}_2 \sum z^2 \end{array} \right.$$

it follows that

$$\hat{\theta} = \hat{\beta}_1 + \hat{\beta}_2 \frac{\sum xz}{\sum x^2}.$$

If (B) is the correct model $\hat{\beta}_1$ and $\hat{\beta}_2$ are unbiased estimators so

$$E(\hat{\theta}) = \beta_1 + \beta_2 \frac{\Sigma xz}{\Sigma x^2}$$

It is evident that if x and z are orthogonal, i.e. such that $\Sigma xz = 0$ then $\hat{\theta}$ is an unbiased estimator of β_1 in the true model and the term in z is completely missed. If, however, x and z are related with $\Sigma xz \neq 0$ then $\hat{\theta}$ is a biased estimator of the coefficient of x in the true model. The bias gets worse as Σxz increases. A further consequence of the misspecification is apparent in the residuals. The residuals from the fitted model (A) are

$$\hat{e}_i = y_i - \hat{\theta} x_i = y_i - \hat{\beta}_1 x_i - \hat{\beta}_2 x_i \frac{\Sigma xz}{\Sigma x^2}.$$

Had we fitted the correct model we would have residuals

$$\hat{u}_i = y_i - \hat{\beta}_1 x_i - \hat{\beta}_2 z_i.$$

Hence

$$\hat{e}_i = \hat{u}_i + \hat{\beta}_2 z_i - \hat{\beta}_2 x_i \frac{\Sigma xz}{\Sigma x^2}.$$

If x and z are orthogonal

$$\hat{e}_i = \hat{u}_i + \hat{\beta}_2 z_i,$$

so the residuals from the fitted model contained the missed variable z_i and a residual plot of $\hat{e}_i$ against z_i would show a linear relationship of slope $\hat{\beta}_2$. This would not be the case, however, if x and z were not orthogonal. A further consequence of this relation is that the estimator of variance, $s^2 = \Sigma \hat{e}^2/(n-1)$, will give an inflated estimate of the underlying variability, which makes it harder to test for the introduction of any new variable.

If x and z are strongly related we could imagine seeking to predict z from x using a subsidiary model $E(z_i) = \alpha x_i$. The predictor of z_i using the least squares estimator of α would be

$$\hat{z}_i = \hat{\alpha} x_i = \frac{\Sigma xz}{\Sigma x^2} \cdot x_i.$$

Thus we could rewrite

$$\hat{e} = \hat{u} + \hat{\beta}_2 (z_i - \hat{z}_i).$$

If $\hat{z}$ is a very good predictor, giving a small error $z_i - \hat{z}_i$, the residual, or prediction error, produced by using model (A) instead of (B) is quite small. So even though we have a biased estimator, we still get good predictions in this case. The reason for this is that x is acting as a proxy variable for z, as well as a variable in its own right. In this case a study of the residuals will not show the influence of z.

If we now reverse the situation and consider the effect of overspecification then model (A), the 'true' model, can be conveniently rewritten as

$$y = \theta x + 0 . z + e.$$

Thus (A) is identical in form to (B) with $\beta_1 = \theta$ and $\beta_2 = 0$. Thus if we estimate by Least Squares $\hat{\beta}_1$ will be an unbiased estimator of θ and $\hat{\beta}_2$ will be an unbiased estimator of zero. The fact that we are using the same data to estimate two β parameters, instead of just one parameter, θ, means that the estimated variance and, consequently, the variance of $\hat{\beta}$, are both inflated.

It is evident that in this last example overspecification caused fewer problems than underspecification. This is part of the reason for the common preference for top-down identification referred to in section 12.2.

Chapter 17

Conceptual, empirical, and eclectic validation

17.1 Conceptual validation

At first sight it may seem wrong to consider validating a model without direct reference to data. There are, however, a number of good reasons for an initial conceptual approach to validation:

(a) The empirical aspects we consider in the next section often provide only limited validation of the overall behaviour of certain variables in the model. Obtaining special data for validation may be a costly task. So our comparisons are often with historical data of limited accuracy and limited coverage, we are in effect doing our validation on one 'observation' of uncertain reliability. Thus conceptual approaches are needed to strengthen the validation.
(b) Conceptual validation can examine individual parameters, variables, and substructures which, as we have already indicated, is often difficult and costly in empirical validation.
(c) Human beings have a great capacity for storing past data in the form of 'feelings' about data and behaviour, intuitive fuzzy models. Though the data may no longer be available or even directly relevant, we are unwise if we neglect this surrogate information. We need to bring this 'expert's opinion' to bear on the validation of the model.

To carry out a conceptual validation it is advisable to start with a systematic checklist of the form of the model and its behaviour. Clearly this checklist must be designed for the specific model. Let us consider some general headings that could be used for such a checklist.

(a) *Model Form.* Does the model show the intuitively correct symmetry or asymmetry?

241

Do the individual components of the model show the right forms of interrelationships?

(b) *Variables.* Do the variables included in the model make sense in causal terms?

(c) *Parameters.* Do the parameters have sensible values? For example, do the dependent variables alter in sign or magnitude in an intuitively correct way when regressor variables are changed?

Are the relative contributions of different terms about right?

How sensitive is the behaviour of the model to errors in parameter estimates?

(d) *Normal Behaviour.* Does the fitted model show reasonable short-term, trending and oscillatory, and long-term, asymptotic, behaviour?

17.2 Empirical validation—parameters

We now turn to the validation of models using data from the prototype. We have termed this empirical validation. This can give the impression of clear objectivity in contrast to the subjectivity of conceptual validation. Such an impression can be misleading, since we rarely know how the real situation actually behaves. We can simply try and measure some aspects of its behaviour. The choice of what to measure and the errors in its measurement can both impact on our validation. Our model may compare well with 'reality' only because we have ignored certain aspects of 'reality'. We may be unaware of this at the identification fitting and validating stages, but it may ruin the application of our model. Even though in this chapter we shall consider a number of formal tests of validity, we must never abandon insight and judgement as the basis for validation.

At several stages in this chapter it will be necessary to distinguish between the use of fitting data, validating data, and working data. Clearly if we compare our fitted model with its fitting data we would expect to get a better fit than would be obtained for validating data or working data. When data is short we may sometimes combine fitting and validating, by fitting with only part of the data and validating with the remainder, the two parts may then be reversed. This procedure is called *cross-validation*.

A further distinction needs to be made between validating a completely fitted model, with all its parameter values numerically given, and validating only the form of the model and re-estimating the parameters from the validation data. Clearly this distinction only has meaning when validation data is being used.

Though we have presented validation as logically distinct from identification, the process of eclectic identification analysis and the process of empirical validation of model form using fitting data are very similar. The practical distinction lies perhaps in the emphasis in identification on informal and graphical techniques and in validation on formal significance tests.

In some situations validation reduces to a testing of model parameters. We examine next three situations in which this can occur.

17.2.1 Testing received parameters

It frequently occurs that a model is a 'received' or standard model that is well supported in the literature, and the parameters are also given as correct values without any indication of the standard errors of their original estimation. There will thus be little doubt about the applicability of the form of the model, but we may wish to check the model parameters.

Example 1
Suppose that before we collected the data, on road distances in Sheffield, of Table 13.1, we were told that the linear relation had proved to be a good model for another city and that $\theta = 1.4$. From the calculations of Chapter 14 we know that, for the Least Squares estimator $\hat{\theta}$,

$$E(\hat{\theta}) = \theta \qquad \text{and} \qquad V(\hat{\theta}) = \sigma^2 / \sum x_i^2.$$

We have as an estimator of σ^2 the unbiased estimator $\hat{\sigma}_e^2 = \Sigma \hat{e}^2 / (n-1)$.

As $\hat{\theta}$ is a linear function of the observations y_i we will have a normal distribution for $\hat{\theta}$ if the errors are normally distributed. With this assumption

$$\hat{\theta} \cap N(1.4, \sigma_e^2 / \sum x_i^2)$$

so

$$\frac{\hat{\theta} - 1.4}{\hat{\sigma}_e^2 / \sqrt{(\sum x_i^2)}} \cap t_{n-1}.$$

The observed value, using the calculations of Table 13.1, is 3.7. The value of t_{19} for a 5% significance test is 2.09 and for 1% is 2.86. Thus the hypothesis that $\theta = 1.4$ is clearly inconsistent with the 20 observations of Table 13.1 and the proposed parameter value is rejected for that data.

17.2.2 Testing model components

A very similar situation occurs where we wish to test a particular component of a model using validation data. In this case we can test whether the parameter associated with that component is zero.

Example 2
In a time series based regression an additive trend and seasonal model could be put in the form

$$x_t = \alpha + \beta t + \gamma \left[\sin\left(\frac{2\pi i}{6} + \phi\right) + \delta \sin\left(\frac{4\pi i}{6} + \psi\right) \right] + e_t,$$

where $i = 1, \ldots, 12$ is the month of the year. A test of the validity of the seasonal term reduces to a test of whether the parameter γ is zero.

17.2.3 Testing and pooling

A somewhat different situation arises when we have a model with parameters whose estimates are known to be rather poor. Here the question is whether the unknown parameter value for the validation data is the same as the unknown parameter value for the original fitting data. If the answer appears to be 'yes' then it would be sensible to pool together the fitting and validating data to get a better estimate. If 'no' then the implications of having different parameter values will have to be worked out for the application proposed.

Example 3

A classic example here concerns the variance of the error term in a model. Suppose the fitted model is

$$y = g(x, \hat{\theta}) + e$$

where the form of the model $g(\ ,\ )$ is known, $\hat{\theta}$ represents a set of estimated parameters and e is taken as Normal $N(0, \sigma^2)$. Suppose the estimate $s_f^2 = 10.6$ of σ^2 is found and is based on $f = 7$ degrees of freedom. Such low degrees of freedom are quite common since most of the information available may have been used in estimating the other parameters. The validation data may now be used to re-estimate σ^2 giving $s_v^2 = 15.1$ with $v = 10$ degrees of freedom. This involved re-estimating θ. Many more degrees of freedom would have been obtained if the $\hat{\theta}$ from the fitted model had been assumed true, but this would have involved an unnecessary and potentially misleading assumption. We saw in Chapter 4 that the distribution of ratios of independent s^2 estimators of the same σ^2 relates to an F distribution by

$$\frac{f s_f^2}{v s_v^2} \cap F_{f,v}$$

so

$$\frac{7 \times 10.6}{10 \times 15} \cap F_{f,v}$$

or

$$\frac{10 \times 15}{7 \times 10.6} \cap F_{v,f},$$

following the convention of statistical tables of putting the larger values in the numerator and consequently using the top $2\frac{1}{2}\%$ critical region for a 5% two-sided test. The $F_{10,7}$ value is 4.76 for $2\frac{1}{2}\%$. As the above observed value is just over 2, there is clearly no evidence to lead us to reject the hypothesis that both data come from the model with the same σ. It is thus reasonable to pool the information from fitting and validating data. The pooled estimate s_p^2 is obtained from

$$s_p^2 = \frac{f s_f^2 + v s_v^2}{f + v}$$

which is just the weighted average of the estimates using the degrees of freedom to provide the weighting. Thus for the above data

$$s_p^2 = \frac{7 \times 10.6 + 10 \times 15.1}{17} = 13.2.$$

17.2.4 Methods of testing

The previous discussion shows the role of parmeter testing in validation. The question raised by this is how to carry out the tests required. There are three commonly used procedures:

(a) Using known distributions

In many standard situations the fitting or validating data can be used to obtain estimators of the parameters which have known distributions and which can thus be used to test parameter values. For example, for a linear model with a normal error distribution, the least squares estimators $\hat{\beta}$ are normally distributed with mean β and a variance that is readily estimated (section 14.3). So the specified β can be tested with a standard t-test as in Example 1.

(b) Using approximate distributions

It often occurs that we do not know the distributions of estimators. Fortunately some of the standard methods of estimation automatically lead to estimators with known approximate distributions for large samples. For example, Maximum Likelihood estimators are aproximately Normally distributed with expectation equal to the true value. Their variance can be found from a quantity called the *Fisher Information*. This is defined by

$$I(\theta) = V\left(\frac{\mathrm{d}L}{\mathrm{d}\theta}\right),$$

where L is the log likelihood. It can be shown that if the maximum likelihood estimator is $\hat{\theta}$, then for large n

$$\hat{\theta} \cap \mathrm{N}(\theta, 1/I(\theta)).$$

As the variance of $\hat{\theta}$ is $1/I(\theta)$ the larger the information the better, the more efficient the maximum likelihood estimator. An alternative expression for $I(\theta)$ that emphasizes this point is

$$I(\theta) = -E\left(\frac{\mathrm{d}^2 L}{\mathrm{d}\theta^2}\right).$$

The second derivative is a measure of curvature so $I(\hat{\theta})$ indicates the expected sharpness of the peak of the log likelihood curve.

Example 4

Observations are taken on a Bernoulli process with probability of success θ, which is thought to be 0.1. The first successes in 10 experiments are at observations $n_i = .6$, 9, etc. where

$$\sum_1^{10} n_i = m = 82.$$

The appropriate model is the Geometric Distribution

$$Pr(N = n) = \theta(1-\theta)^{n-1}, \quad n = 1, 2, \ldots,$$

which has expectation $1/\theta$ and variance $(1-\theta)/\theta^2$.

With k observations, n_i, the log likelihood is

$$L(\theta) = k \ln \theta + (m-k) \ln(1-\theta), \text{ where } m = \sum_{i=1}^k n_i,$$

so

$$\frac{dL}{d\theta} = \frac{k}{\theta} - \frac{m-k}{1-\theta}$$

This quantity, which is a random variable, is the *score* denoted by $U(\theta)$.

Remembering that the n_i are k independent random variables, it will be seen that

$$E\left(\frac{dL}{d\theta}\right) = \frac{k}{\theta} - [kE(n) - k]/(1-\theta) = 0,$$

and

$$V\left(\frac{dL}{d\theta}\right) = \frac{k}{(1-\theta)^2} V(n) = \frac{k}{\theta^2(1-\theta)}.$$

It follows that the null hypothesis is

$$H_0: \theta = 0.1 \text{ and under } H_0 \; 1/I(\theta) = 0.0009.$$

Thus $\hat{\theta}$ is distributed approximately as $N(0.1, 0.0009)$.

The acceptance region for a 5% significance test is

$$(0.1 \pm 1.96\sqrt{0.0009}) = (0.041, 0.159)$$

The maximum likelihood comes from $L(\hat{\theta}) = 0$ which gives $\hat{\theta} = k/m$, which is $\hat{\theta} = 0.12$ and hence we accept the null hypothesis and conclude that the data is consistent with $\theta = 0.1$.

An alternative approximate test is based on the approximation, for large n, that the score, $dL/d\theta$, is itself normally distributed with mean zero and variance $I(\theta)$.

Example 5

Returning to the geometric example, the observed score, u, under H_0, given $m = 82$, is

$$u = \frac{10}{0.1} - \frac{(82 - 10)}{0.9} = 20$$

The standard deviation for u is $\sqrt{I(\theta)} = 33.3$ so the critical region is that outside ($\pm 1.96 \times 33.3$). The observed u is well within the central acceptance region so the observation is consistent with H_0.

Thus there are two approximate tests of a hypothetical parameter value based on the likelihood. Of these the one based on using the score has the merit that it does not require the maximum likelihood estimate to be known.

(c) Using simulated distributions

In the worst case we may know how the estimator was obtained but little about its properties other than its numerical value. The validation data can be used to obtain a single estimator $\hat{\beta}_v$. A useful procedure in this situation is to make use of simulation via a *Monte Carlo Significance Test*. Given a complete fitted model it is always possible to simulate sets of data corresponding to the validation data. If there are n validation observations then N samples of n simulated observations could be obtained. Each of these could be used to obtain an estimator giving $\hat{\beta}_1, \ldots, \hat{\beta}_N$. We now add $\hat{\beta}_v$ to this sample and put in order the $N + 1$ estimators. If the validation data supports the numerical values used in the fitted model, then we would expect $\hat{\beta}_v$ to lie inconspicuously in with the other estimators. If it is at either end, then the simulation mechanism is not consistent with the validation data. The basic rule is to reject consistency at the α significance level if $\hat{\beta}_v$ is in one of the $N\alpha$ most extreme positions. Note that this test tests $\hat{\beta}_v$ against the whole fitted model and not just the fitted parameter. However, an exploration of all parameters in this fashion may suggest where things are wrong. An alternative, where there is extensive validation data and the estimated parameters can be regarded as approximately independent, would be to test β as explained above, but to use in the model estimates of all other parameter obtained from the validation data rather than the fitting data.

Before leaving this section it is worth emphasizing the need to distinguish between statistical significance and technical significance. Usually we are short of data and if we reject a hypothesized parameter value, θ, it will probably be because the value for the validation data is not only statistically significantly far from θ but technically far in the sense of making a major difference in our model use. An understanding of what constitutes a technically significant change is a product of a careful sensitivity analysis. However, if a large amount of data is available we shall be able to detect much smaller variation in parameter values. So we may reject a θ, on detecting a statistically significant difference, which in fact is too small to worry us as a technically significant difference. This can be examined by using the newly estimated $\hat{\theta}$ in a sensitivity analysis of the model.

In general it needs emphasizing that the results of significance tests are not prepackaged decisions. They provide evidence of possible consistency or inconsistency between data and hypotheses about models. This evidence needs thinking about.

17.3 Empirical validation—form

When we estimate model parameters we seek directly or indirectly a good fit to the model; thus at the validation stage a lack of fit will indicate a wrong choice of model form. We concentrate in this section on various ways of testing the quality of fit of the model. The approaches to doing this vary with type of model under consideration.

17.3.1 Testing fit for distributional models

The basic question in validating distribution models is whether the shape of the fitted model corresponds to that of the data. To do this we may simply make a direct comparison of the observed data with what we expect to see from the fitted model. Thus if we have data in a frequency table the fitted model, $\widehat{Pr}(r)$, can be used to estimate the probabilities, $Pr(r)$. Multiplying by the number of observations, n, gives the expected frequencies

$$\hat{f}_r = n\,\widehat{Pr}(r).$$

These can be compared directly with the observed frequencies f_r.

It is useful to put this informal procedure in the form of a test. To do this we make use of the fact, proved in the more advanced texts in statistics, e.g. Kendall and Stuart, that the statistic

$$X^2 = \sum_{r=1}^{k} \frac{(f_r - \hat{f}_r)^2}{\hat{f}_r}$$

has approximately a χ^2-distribution. This result applies to any situation where f_r is an observed frequency and $\hat{f}_r$ is the corresponding expected frequency based on some model. The degrees of freedom are:

Number of frequencies $-1-$ number of parameters
estimated with the same data.

The adjusted for the number of estimated parameters will not be needed if validating data is used with a model having parameters previously estimated with fitting data. The one is subtracted since the $\hat{f}_r$ are constrained to give $\Sigma \hat{f}_r = \Sigma f_r$. To get a good approximation n needs to be large and values of $\hat{f}_r$ should not be small, not much less than 5. This procedure is usually called the χ^2 *Goodness of Fit Test*.

Example 1

The use of Poisson Probability Paper in Chapter 11 suggested a Poisson model was suitable for some data based on the movement of leaf hoppers across a sand dune. Table 17.1 shows the calculation of the χ^2 goodness of fit test, which confirms that the data is consistent with the Poisson model.

Table 17.1 Goodness of fit of Poisson Model

(a) Data

Hemiptera (leaf hoppers)

Individuals per trap r	Frequency f_r	rf_r	$\widehat{Pr}(r)$	$\hat{f}_r$	$\dfrac{(f_r - \hat{f}_r)^2}{\hat{f}_r}$
0	6	0	0.183	6.04	0.00
1	8	8	0.311	10.26	0.50
2	12	24	0.264	8.71	1.24
3	4	12	0.149	4.92	0.17
4 +	3	12	0.093	3.07	0.00
	33	56	1.000	33.00	1.91

$$\hat{\mu} = 56/33 = 1.7 \text{ (as } f_5 \text{ etc.} = 0)$$

$$e^{-\hat{\mu}} = 0.183$$

$$\widehat{Pr}(r) = e^{-\hat{\mu}} \frac{\hat{\mu}^r}{r!} \qquad \hat{f}_r = 33\,\widehat{Pr}(r)$$

$$\sum \frac{(f_r - \hat{f}_r)^2}{\hat{f}_r} = 1.91 \qquad \chi^2_{5-2} = \chi^2_3$$

The value 1.91 is less than the mean of χ^2_3 which is 3 and lies in the central region of the distribution. Thus the data is consistent with the Poisson model.

An alternative test formalizes the use that was made of cumulative distribution curves in model-based identification. In section 11.2 we made use of several plots that compared the theoretical cumulative distribution function $F(x)$ with the empirical c.d.f. $\hat{F}(x)$. If the model is valid $\hat{F}(x)$ should not depart too much from $F(x)$. An appropriate test statistic is the maximum difference between these two, i.e.

$$D_n = \underset{x}{\text{maximum}} |\hat{F}(x) - F(x)|.$$

This is called the Kolmogorov–Smirnov statistic. If D_n exceeds a tabulated value d_n, which depends only on the number of observations, n, then we reject the hypothesis that the model $F(x)$ is correct. Tables of d_n are given

Table 17.2 Kolmogorov–Smirnov test of the data of Table 17.1

Individuals per trap	f_r	Cumulative frequency	$\hat{F}(x)$	$F(x)$	$|\hat{F}(x) - F(x)|$
0	6	6	0.182	0.183	0.001
1	8	14	0.424	0.494	0.070
2	12	26	0.788	0.758	0.030
3	4	30	0.909	0.907	0.002
4	3	33	1.00	1.00	0.000

$\hat{F}(x)$ is the cumulative frequency/33.
$F(x)$ estimated from cumulative $\hat{Pr}(r)$ of Table 17.1
$D_n = 0.07$ $\qquad n = 33$
From tables, e.g. Neave, for 5% significance, D_n must exceed 0.23. Thus the hypothesis of the Poisson model is accepted.

in, for example, Neave (see references for Part II). Table 17.2 illustrates the method. This test leads to a useful by-product for at the α level of significance we can state the test as

$$Pr\{D_n > d_{n,\alpha}\} = \alpha.$$

Substituting for D_n and re-organizing gives

$$Pr\{\hat{F}(x) - d_{n,\alpha} \leqslant F(x) \leqslant \hat{F}(x) + d_{n,\alpha}\} = 1 - \alpha.$$

This probability statement is true for all x and so it presents a confidence statement in which the true $F(x)$ is sandwiched between the empirical curves $\hat{F}(x) - d$ and $\hat{F}(x) + d$. Such a 'confidence band' is helpful in identification by putting some limitation on possible $F(x)$ to be considered.

For models with only one parameter both the mean and variance must depend on that parameter and hence there must be a relation between mean and variance.

For example, we have

Poisson $P(\mu)$ $\qquad E(R) = \mu,$ $\qquad V(R) = \mu,$ $\qquad$ so $V(R) = E(R)$
Binomial $B(n, \theta)$ $\qquad E(R) = n\theta,$ $\qquad V(R) = n\theta(1 - \theta),$
$\qquad\qquad\qquad\qquad$ so $V(R) = E(R)(1 - E(R)/n).$

Given a large sample of m observations on each of these distributions the sample variance has approximately a χ^2 distribution with $m - 1$ degrees of freedom, i.e.

$$(m - 1)s^2/\sigma^2 \cap \chi^2.$$

The population variance, σ^2, can be approximated using the sample mean and the relation between variance and mean so that we have

Poisson: $\qquad (m - 1)s^2/\bar{x} \cap \chi^2$

Binomial: $\qquad (m - 1)s^2/\bar{r}\left(1 - \dfrac{\bar{r}}{n}\right) \cap \chi^2.$

These statistics thus provide simple tests of the form of the distributions. Such tests are called *Dispersion* tests.

Example 2

Returning to the leaf-hopper data of Table 11.1 we have

$$\bar{x} = 1.7, \ s^2 = 1.4, \ m = 33$$

so

$$(m-1)s^2/\bar{x} = 26.5.$$

This is in the central region of the χ^2_{32} distribution and hence we confirm that the data is consistent with the Poisson model. The analogous calculation for the rove beetle data of Table 11.1 gives a value of 87.6 which is not consistent with the Poisson assumption.

17.4 Testing form—regression models

In section 12.3 we discussed briefly a range of methods of testing variables in a regression model. The objective at that stage was to identify a suitable model. Many of the ideas of that chapter also have relevance in validation. The importance of a variable in the regression is a consideration in both identification and validation. At the validation stage a more general consideration will also be appropriate, namely whether the whole proposed regression model contributes anything to our ability to describe the prototype.

17.4.1 Testing with fitting data

Consider the simplest regression line, through the origin, $y_i = \theta x_i + e_i$. Given the fitted model $\hat{y} = \hat{\theta}x$ we have

$$y_i = \hat{y}_i + \hat{e}_i$$

which breaks the fitting data into contributions from the fitted model and from the residuals. Squaring and summing gives

$$\sum y^2 = \sum (\hat{y} + \hat{e})^2$$
$$= \sum \hat{y}^2 + 2\sum \hat{y}\hat{e} + \sum \hat{e}^2$$
$$= \sum \hat{y}^2 + 2\hat{\theta}\sum x\hat{e} + \sum \hat{e}^2,$$

but the normal equations for the regression model can be expressed as $\Sigma x\hat{e} = 0$, Example 4, section 13.3. So

$$\sum y^2 = \sum \hat{y}^2 + \sum \hat{e}^2$$

If we relax the condition that the line goes through the origin, as in this example, the model can be expressed in the form

$$y_i - \bar{y} = \theta(x_i - \bar{x}) + e_i$$

and the above equation becomes

$$\sum (y_i - \bar{y})^2 = \sum (\hat{y}_i - \bar{y})^2 + \sum \hat{e}_i^2.$$

Thus the total sample variability of y, which is measured by $\sum (y_i - \bar{y})^2$ can also be broken into contributions from the fitted model and the residual. The term $\sum (\hat{y}_i - \bar{y})^2$ gives a measure of the variability of the $\hat{y}_i$ about their average, since $\sum \hat{e}_i = 0$ implies that $\sum \hat{y}_i = \sum y_i$, so y_i and $\hat{y}_i$ have the same average. The term $\sum \hat{e}_i^2$ gives the measure of variability of the observations about the fitted line. The relationship can be expressed in a tabular form as in Table 17.3. The mean square column is obtained by dividing the sums of squares by their degrees of freedom. Suppose the model contributes nothing to the explanation of y, i.e. all its parameters are zero, and also that the errors are normally distributed. Then in this case s_m^2 and s_r^2 are independent estimators of the underlying variance, σ^2, and thus their ratio has an F distribution.

Table 17.3 Analysis of variance for a regression model with parameters β_0, $\beta_1, \ldots, \beta_k$

Source of variation	Sum of squares (1)	Degrees of freedom (2)	Mean square (1)/(2)	Ratio
Due to model	$\sum (\hat{y}_i - \bar{y})^2$	k	s_m^2	
Residual	$\sum \hat{e}_i^2$	$n - 1 - k$	s_r^2	$F = \dfrac{s_m^2}{s_r^2}$
Total	$\sum (y_i - \bar{y})^2$	$n - 1$		

We can thus use this F test as a test of the complete model. Table 17.4 illustrates the method. If the model using the pre-test was helpful in explaining the final test marks, the sum of squares due to the model would make a larger contribution to the total sum of squares, the residual sum of squares would be smaller, and the F value would be significantly large. The method is of value if the random element tends to obscure any deterministic structure. The method of breaking the total variation into various contributory sources is called the *Analysis of Variance*.

17.4.2 Testing with validation data

It has been frequently emphasized that the residuals contain all information that is not explicitly in the model. Thus formal tests using the residuals provide a natural means of validating models. If we wish to test a complete fitted regression model, with normal errors, using validation

Table 17.4 Example of analysis of variance of a regression model

Marks from a course test (y) and the course pre-test (x) are obtained from 20 students. The predictive model is $y = \beta_0 + \beta_1 x + e$ which is fitted as

$$\hat{y} - \bar{y} = \hat{\beta}_i (x - \bar{x}) \qquad \text{(Example 4, section 13.3)}$$

The data gives

$$\sum x = 1040, \qquad \sum x^2 = 58200, \qquad \sum xy = 61422.$$
$$\sum y = 1160, \qquad \sum y^2 = 72148$$

To calculate the quantities of Table 17.3 we use

$$\sum (y - \bar{y})^2 = \sum y^2 - \frac{(\sum y)^2}{n} = 4878$$

$$\hat{\beta}_1 = \left[\sum (x - \bar{x})(y - \bar{y}) \Big/ \sum (x - \bar{x})^2 \right] = \left[\sum xy - \frac{\sum x \sum y}{n} \right] \Big/ \left[\sum x^2 - \frac{(\sum x)^2}{n} \right]$$

$$= 1102/4120 = 0.267$$

$$\sum (\hat{y} - \bar{y})^2 = \hat{\beta}_1^2 \sum (x - \bar{x})^2 = 294.758$$

$\Sigma \hat{e}^2$ is obtained by subtraction in the table.

Source of variation	Sum of squares	Degrees of freedom	Mean square	Ratio
Due to model	294.8	1	294.8	1.16
Residual	4583.2	18	254.6	
Total	4878.0	19		

An F value of 1.16 is near the centre of the F-distribution. Thus there is no evidence that the regression model is helpful in explaining the course marks y.

data (x_i, y_i) we have

$$y_i = f(\hat{\beta}, x_i) + e_i \qquad \text{fitted model, } \hat{\beta} \text{ given.}$$
$$\hat{y}_i = f(\hat{\beta}, x_i) \qquad \text{predicted } y\text{'s.}$$

The residuals are $y_i - \hat{y}_i = e_i$ which are the true model errors, if the model is correct. So $y_i - \hat{y}_i$ which is determined from the data actually equals the e_i which are independent $N(0, \sigma^2)$ under the assumed model. Each of these four standard regression assumptions (independence, zero expectation, variance σ^2, normality) can then be tested. Unlike the previous methods this approach does not require the regression to be linear. We may use any functional form that will give $\hat{y}$ in terms of the independent variables and parameters, which are known. In this situation we can also handle multiplicative errors since where these exist we have

$$y_i / \hat{y}_i = e_i$$

so again the assumptions about e and, by implication, about the complete fitted model may be tested. Where the fitted model provides an estimate s_0^2 of σ^2 based on known degrees of freedom, v, then given n errors e_i we have

$$F = \frac{\sum e_i^2/n}{s_0^2/v}$$

as a test statistic with an F-distribution with n, v degrees of freedom. This will lead to rejection of the model if either $E(e) = 0$ or $V(e) = \sigma^2$ are not valid assumptions.

Where only fitting data is available we can move some way towards using the above approaches by *Jackknife Validation*. In this we first drop the ith observation and then estimate the parameters, fit and predict y_i giving $\hat{y}_{-i}$. The estimated error is then $\hat{e}_i = y_i - \hat{y}_{-i}$. The properties of $\hat{e}_i$ are much closer to those of e_i above than the usual residual. Suppose the observations arise as a time series and $\hat{y}_t$ is the predicted value of y_t, using the estimate, $\tilde{\beta}_{t-1}$ obtained from the data up to time $t-1$, and the new regressor variable x_t, then the quantity

$$\tilde{e}_t = y_t - \hat{y}_t = y_t - f(\tilde{\beta}_{t-1}, x_t),$$

is called a *Recursive Residual*. The sequence of recursive residuals again have the property of zero mean and independence and consequently can be conveniently used to test model validity.

17.5 Testing form—progression models*

Techniques of residual analysis apply equally to the estimates of the innovations in progression models. Thus the $\hat{e}_i$ might arise from a fitted $AR(p)$, $MA(q)$ or a $D(p, q)$ model. To obtain these residuals we need an initial identification of the form of the model. For the ARMA class of models this can be done by examining the autocorrelations and related quantities, since:-

(i) For a $MA(q)$ process
The autocorrelations, ρ_h are zero for $h > q$. The partial autocorrelations, ϕ_{hh}, form a generally non-zero infinite sequence of values.
(ii) For the $AR(p)$ process
The partial autocorrelations, the coefficients, ϕ_{hh}, are zero for $h > p$. The autocorrelations, ρ_h, form a generally non-zero infinite sequence of values.

Hence we first obtain estimates of ρ_h and ϕ_{hh}, for $h = 1, 2, 3, \ldots$. If the model is either one or the other of the MA or AR families the tendency of one to cut off suddenly to values close to zero and the other to remain non-zero as h increases will indicate which of the two models is appropriate. The value of the cut-off point, if clear, will indicate the order, p or q, of the process. Figure 17.1 shows some illustrative results. The technical details

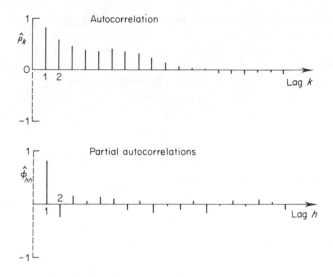

Figure 17.1 Plots of ρ_k and ϕ_{hh}. These two plots suggest that an MA process is not appropriate but an AR (1) might be

of obtaining the estimates of ϕ_{hh} are beyond the scope of this book; suffice it to say that computer programs for the calculations are commonly available but require substantial amounts of data (more than 50 observations) to give reliable results (see Box and Jenkins (in references for Part II) and Chatfield (references for Part VI).

Having used this graphical method to obtain an initial identification then the residuals from the models can be analysed to check this identification. All the progression models are based on interrelations through time in which the basic randomness comes from the innovations, e_i, which are a sequence of independent random $(0, \sigma^2)$ quantities with a constant form of distribution. Estimates of these are obtained as the one step ahead prediction errors, see, for example, Example 3 of section 18.4. The quality of the model is best tested by examining whether these are independent. Denote the autocorrelations of these by $r_{e1}, r_{e2}, r_{e3} \ldots$ for the lags 1, 2, 3, etc. They can all be examined separately but it is convenient to have a single test that all the reasonably lagged autocorrelations are zero. This is provided by *Portamanteau tests*; the most common version of this uses the test statistic

$$Q = n \sum_{i=1}^{P} r_{ei}^2,$$

where n is the number of observations and P is the number of autocorrelations used, which should be large enough to include all those of possible significance from reasonable alternative models. The distribution of Q is approximately χ_D^2 if the errors are independent. Thus the value of Q is found and the model accepted if Q is consistent with the χ_p^2 distribution.

For regression and progression models in general an important test of the validity of the model is whether the residuals or prediction errors form an independent series. A more common test for independence of adjacent values is the *Durbin–Watson Test*. The test statistic is

$$d = \frac{\sum\limits_{t=2}^{n} (\hat{e}_t - \hat{e}_{t-1})^2}{\sum\limits_{t=1}^{n} \hat{e}_t^2}.$$

If we write the estimated first order autocorrelation by

$$r_1 = \frac{\sum \hat{e}_t \hat{e}_{t-1}}{\sum \hat{e}_{t-1}^2}.$$

A little algebra will show that approximately

$$d = 2(1 - r_1).$$

So for series that are independent, or at least uncorrelated for lag one, the statistic d will be about 2. For high positive or negative correlation it will tend to zero or 4. Separate tests are used to test against the alternative hypotheses of positive or negative correlations. Tables for the test are given in many books of tables, e.g. Neave.

17.6 Validating a working model

As we live in an ever changing world it is unwise to assume a model that was satisfactorily validated last year will still be satisfactory next year. If we are to use a model in a working situation over a long period, it is necessary to consider how the validation process can be continued over that period. We may refer to this continuous validation as a monitoring process. Setting up such a process requires an examination of the ways in which the fitted model might become unsatisfactory. For example, in some situations a model is initially identified and fitted using data from a large initial study. The model may then be used in a situation where only limited data is available. In such a situation a change in a parameter value may invalidate the whole model as data is not available to re-estimate its value. In other situations a parameter may be estimated in such a fashion that it is always adjusting to new information, so the fitted model simply adapts to changes in parameter values and remains valid in spite of the change in parameter value. Thought needs to be given to the types and magnitudes of changes in model structure and fit that would lead to a loss of validity in relation to the model's particular application. For example, in using a model for prediction, a change in mean may be more significant than a change in slope if short-term predictions are being made, but less significant if long-term predictions are important.

The monitoring of working models is a large and relatively unexplored topic. There are at least as many approaches as there are types of model. The aim of this section is simply to illustrate a few of these.

The simplest and historically the oldest approaches to monitoring are the classical quality control techniques. These were developed to monitor fitted distributional models occurring in production processes. For example, a normal distribution $N(\mu, \sigma^2)$ may provide a valid model for the resistances of a mass-produced resistor when the production process is under proper control. For this model 95 % of resistances will be between approximately $\mu - 2.\sigma$ and $\mu + 2.\sigma$ ohms, $2\frac{1}{2}\%$ less than $\mu - 2.\sigma$ ohms and $2\frac{1}{2}\%$ greater than $\mu + 2.\sigma$ ohms. If the underlying structure changes the proportion of resistors lying in the three zones will change. In practice the process is usually monitored by taking, not all resistors but, samples and using the mean resistance $\overline{X}$ as $N(\mu, \sigma^2/n)$ to provide the basis for a *Control Chart,* as in Figure 17.2. In this chart two additional zones are added for further sensitivity.

When we turn to regression and progression models the natural bases for monitoring model validity are the recursive residuals or prediction errors. If the fitted model remains valid then, given regressor variables and/or past values, the dependent variable y_t can be predicted from the model as $\hat{y}_t$. The prediction error is then

$$\hat{e}_t = y_t - \hat{y}_t$$

If the model remains valid the sequence of the $\hat{e}_t$ will form a random sequence with zero mean and constant variance. This can be tested either informally by simple plotting $\hat{e}_t$, or rather more formally using appropriate significance tests on samples of data, for example the Durbin–Watson Test.

A third approach to monitoring can be used when our initial studies suggest that, if a model becomes invalid, then the change in structure will occur in certain predictable ways. Such studies will point to particular parameters, estimators, or other characteristics of the data that should be

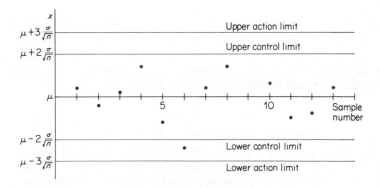

Figure 17.2 A control chart

258

watched. This knowledge may be used to construct from the working data
an indicator of the behaviour of particular interest. Suppose we have a set
of prediction errors, $\hat{e}_i$, then we could construct the following indicators

(i) indicators of mean $\quad I_{m,i} = \hat{e}_i$ or $I_{s,i} = $ sign of $\hat{e}_i$
(ii) indicators of variability $\quad I_{v,i} = \hat{e}_i^2$ or $I_{ab,i} = |\hat{e}_i|$
(iii) indicators of autocorrelation $\quad I_{a,i} = \hat{e}_i\hat{e}_{i-1}$ or $I_{c,i} = \hat{e}_i - \hat{e}_{i-1}$
(iv) indicators of cost $\quad I_{c,i} = $ cost $(\hat{e}_i)$

The validity of the model can be monitored by plotting the indicators as
they are. However, the large randomness in many situations tends to hide
small but significant underlying changes. To reduce the effect of randomness it is helpful to smooth out the indicators by a moving or weighted
average. A common form of smoothing uses a geometric or exponential
form of weighted average. Thus for any of the indicators, I, the smoothed
indicator at time t is given by

$$\tilde{I}_t = S_t/W_t,$$

where

$$S_t = \sum_{r=0}^{t-1} a^r I_{t-r} = I_t + aI_{t-1} + a^2 I_{t-2} + \ldots + a^{t-1}I_1$$

and

$$W_t = \sum_{r=0}^{t-1} a^r \quad = 1 + a \quad + a^2 \quad + \ldots + a^{t-1},$$

where a is a *Smoothing Constant*, usually chosen in the range (0.8, 0.95).
This can be easily calculated by repeatedly using the following relations as
new data arises:
$$S_t = I_t + aS_{t-1}, \quad S_0 = 0$$
$$W_t = 1 + aW_{t-1}, \quad W_0 = 0.$$

Figure 17.3 shows such smoothed values and Table 17.5 illustrates their
calculation.

One problem with some of the above smoothed indicators is the lack of a
scale upon which to judge their magnitude. In some cases the smoothed
value may be compared with the predicted value of the indicator based on
the fitted model. Thus if a fitted regression has an estimated error variance
of σ^2 then a natural indicator of variability would be

$$T_{v,t} = \tilde{I}_{v,t}/\sigma^2$$

This will be numerically close to one if all is well.

If the objective is to assess the numerical value of the smoothed
indicator of bias, $\tilde{I}_{m,t}$, we could divide it by some indicator of spread, as we
do when we use $(\bar{x} - \mu)/(\sigma/\sqrt{n})$ to standardize the mean $\bar{x}$, which is $\tilde{I}_{m,t}$ for a
$= 1$. Instead of using $\sigma/\sqrt{n}$ as the measure of spread we use $\tilde{I}_{ab,i}$, the

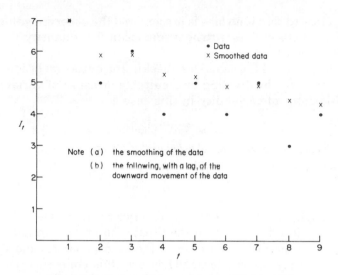

Figure 17.3 Smoothing of data of Table 17.4

Table 17.5 Tabular layout for an exponentially weighted average

(1)	(2)	(3)	(4)	(5)	(6)	(7)
			$(2)+(3)$		$1+(5)$	$(4)/(6)$
Time	I_t	$0.8\,S_{t-1}$	S_t	$0.8\,W_{t-1}$	W_t	$\tilde{I}_t$
1	7	0.0	7	0.0	1.0	7.00
2	5	5.6	10.6	0.8	1.8	5.89
3	6	8.48	14.48	1.44	2.44	5.93
4	4	11.58	15.58	1.95	2.95	5.28
5	5	12.48	17.47	2.36	3.36	5.20
6	4	13.97	17.97	2.69	3.69	4.87
7	5	14.38	19.38	2.95	3.95	4.91
8	3	15.50	18.50	3.16	4.16	4.45
9	4	14.80	18.80	3.33	4.33	4.34

smoothed absolute error. The measure of bias is thus

$$T_{b,\,t} = \tilde{I}_{m,\,t}/\tilde{I}_{ab,\,t}.$$

Relating these to $\tilde{I} = S_t/W_t$ gives

$$T_{b,\,t} = S_{m,\,t}/S_{ab,\,t},$$

since W_t cancels. If we consider the two extremes when the $\hat{e}_t$ are either all positive or all negative then $T_{b,\,t} = +1$ or -1 respectively. Thus in general

$$-1 \leqslant T_{b,\,t} \leqslant 1$$

with $T_{b, t}$ close to zero if no bias is present and the observed values scatter about the predicted values with near zero mean. The quantity $T_{b, t}$ is called a *Tracking Signal for Bias*.

An analogous tracking signal for checking on autocorrelation of lag one is formed from $\tilde{I}_{a, t}$ by dividing by a suitable measure of variability. The natural indicator of variability in this case is

$$I_{d, i} = |\hat{e}_i||\hat{e}_{i-1}|.$$

Thus the *Tracking Signal for Autocorrelation is*

$$T_{a, t} = \tilde{I}_{a, t}/\tilde{I}_{d, t}$$
$$= S_{a, t}/S_{d, t}.$$

A little thought will show that this also lies in the range $-1 \leqslant T_{a, t} \leqslant 1$ with a value close to zero when the prediction errors are independent.

All the above quantities are designed to follow changes and show them clearly when they become large. They are thus concerned particularly with evolving situations. If it is suspected that a sudden change, a fracture in the structure, might occur a slightly different approach is called for. Fractures are best detected by comparing indicators derived before and after a possible fracture. It would be hoped to detect a fracture soon after it has occurred so in practice there will be little data available after the fracture. Thus weighting is inappropriate and we will simply average the indicators obtained after some past time, t, giving $\overline{I}_t$. The smoothed indicator $\tilde{I}_t$ refers to times up to t. Hence

$$c_t = \overline{I}_t - \tilde{I}_t$$

is a measure of any change taking place after t. It should vary about zero except where a fracture occurs.

If I toss a coin ten times and obtain all heads I would rightly suspect the validity of the model of an unbiased coin with $Pr(\text{Head}) = 0.5$. If, however, I do a very long series of several thousands of tosses, with an unbiased coin, then I would be surprised if such a sequence of ten heads did not occur, since its probability is only $(\frac{1}{2})^{10} = 1/1024$. This point needs bearing in mind if any of the methods of monitoring models are to be suitably used. Given a sufficiently long period of use all sorts of extreme behaviour will be observed. Its occurrence may indicate the invalidity of the model, but alternatively it may be a quite natural phenomena. Thus the techniques in this section should not be regarded as decision rules leading to clear choices of validity or invalidity. They are designed to draw attention to, to indicate, possible problem areas. They are the empirical tools, designed to lead on to a conceptual validation, to a careful consideration of what may be happening to the prototype that might invalidate the model.

The title of this chapter referred to eclectic validation yet we have considered only conceptual validation, briefly, and empirical validation. The very nature of validation emphasizes the empirical approach to modelling. Nonetheless there are eclectic aspects that bring together empirical and conceptual considerations. In particular there is the conceptual consideration of the possible alternative models, if the proposed model is found to be, or becomes, invalid. We mentioned this earlier in this section in relation to the choice of indicator. This consideration also has relevance in the choice of tests, referred to in previous sections. We have only considered fairly general tests of validity. Sometimes, however, conceptual considerations suggest that a specific model would be the natural alternative to the one proposed. In these cases we can usually find more effective tests of validity, for example by using some of the ideas discussed in Chapter 12, for choosing between alternative models.

Part VI

Application

Chapter 18

Application

18.1 Introduction

A fascinating activity for anyone interested in modelling is to explore how those concerned with different areas of human interest, Biologists, Engineers, Psychologists, Historians, etc., have made use of models and how particular models, for example the simple linear regression model, keep reappearing in different guises. A survey of applications along these lines would unfortunately take up several volumes and is not directly relevant to the theme of this book. Our concern with 'applications' is with the place of the application of the model in the context of the whole process of statistical modelling. We are thus concerned with how the nature of the application may be taken into account when carrying out the processes of identification, estimation, and validation. For convenience of presentation Application is Part VI of this text. However, in practice it must be part of the very first consideration. The model is developed to help solve a problem. The identification and elucidation of this problem is a necessary preliminary to the modelling process. The nature of the problem will then influence the attitudes and methods of the modeller at the various stages of the modelling process. As an aid to illustrating some of these attitudes and methods we have classified problems as problems of *Description*, *Exploration*, *Prediction*, and *Decision*, covered in sections 18.2, 18.3, 18.4 and 18.5 respectively. In section 18.6 we examine the links between application and modelling in more general terms.

18.2 Description

All the models in this book are at the very least used as descriptions of the behaviour of a prototype. Often the model is created for purposes other than simple description but the validation process will require that it gives a reasonable description of the important features of the prototype. In some applications such a description is enough. The model is used simply as a means of producing a concise description and of summarizing a possibly large amount of data arising from the situation.

Example 1
The easterly position, x, of an aircraft being tracked is described by $x_t \cap N(\hat{\mu}_t, \hat{\sigma}_t^2)$, where the tracking system continuously updates the estimators $\hat{\mu}_t$ and $\hat{\sigma}_t^2$.

Example 2
As part of a large simulation model of a factory layout a particular work station besides a conveyor belt is described in terms of:

(a) items arriving at random, with times between arrival exponentially distributed with mean time 40 seconds;
(b) an exponentially distributed processing time for each item with mean 26 seconds.

In these examples the model acts as an empirical, numerical summary of clearly defined evidence. What is referred to as a law, for example in physics, is a similar summary, but it is based on a long history of empirical evidence that is built up via many different experiments and observations. For example, the inverse square law of gravity provides such a description. Such a law, however, goes beyond any actual body of experimental data and is intended to provide a description of any future evidence. It thus has a predictive role and also it is seen as a conceptual description and not just an empirical one. It is nonetheless a description—it says what happens and not why it happens. Such a description may turn out to be wrong, inadequate, or perhaps valid only in limited circumstances.

In complex systems, such as those modelled by systems dynamics models, the model seeks to include in its description all those features of the prototype that are thought to be significant. This often implies that different parts of the model use different levels of description. Some parts of the model will involve detailed cause and effect relationships. It might be sufficient to describe relatively minor variables in the model, for example, in terms of a simple time trend, without seeking to introduce variables that might explain such a trend

An important aspect of descriptive models is their use as a means of communication. The structure and variables in the model make the modellers approach to the prototype quite explicit.

Example 3
The Global Models produced in recent years were primarily developed for prediction purposes, but they prompted a major debate on how to describe the world system, on what variables are important, on what relations exist between them. They presented clearly and logically a structure that could be analysed and debated.

A major application of the descriptive aspect of models is in the testing of theory. A theory can often be used as a basis for a conceptual model, that can then be fitted. The descriptive validity of the model then provides an indication of the consistency of the theory and the empirical evidence. A total inconsistency may invalidate the theory. It may be that several different theories lead to either the same model or different models which are equally consistent with the data. So this approach may falsify some theories but it will not prove the truth of others.

Example 4
The reference by Zinnes gives an analysis of thirteen different theories, proposed by Richardson, of the involvement of states in wars. The central quantity of this analysis is the probability, $p(r, s)$, that a war will be between r states on one side and s on the other. The various theories, for example that wars are a result of chance, independent, decisions of states, or that there is a tendency to always join the winning side, are then used to develop models for $p(r, s)$. The reasonableness of the models and their consistency with the data is then used as the basis for a critique of the theories.

Having illustrated the descriptive use of models we turn now to the effect of this type of use on the major processes of statistical modelling.

(a) *Identification*: For descriptive model use we emphasize the empirical aspects of identification, since these are based on finding models that describe the evidence of observation. We shall not worry too much about having a theoretical justification for the model and may well use purely empirical models such as those discussed in section 3.5.
(b) *Estimation*: Methods of shape equalization and minimum deviation are natural approaches to estimation for descriptive applications. From the conceptual viewpoint estimators that are unbiased and have small variance would also appear as natural for descriptive purposes.
(c) *Validation*. The natural emphasis here must be how well the normal behaviour of the model matches that of the prototype and also on the descriptive quality of the model when comparing observed with the expected behaviour.

18.3 Exploration

The second class of applications we have termed 'exploration'. In many situations we would like to explore how the prototype behaviour changes if input variables behave differently, if parameter values change, or internal structures were modified. To make such changes to the prototype are often impracticable, so a model is constructed and the exploration is conducted on the model instead of the prototype. For example large economic models are used to investigate the possible effects of budget changes, large urban models are used to study the effects of alternative policies of positioning industrial buildings and public housing. So exploration is essentially based on seeing the model as conditional on various assumptions and on seeking to answer 'what if . . . ?' questions about the effects of changing these assumptions.

A further group of applications that can be included under the 'exploration' heading are those concerned with design. Usually in design problems a number of models are brought together to produce an overall model or system having some desired properties. For example the factory model referred to in the previous section could be used to explore the effects of different numbers of workstations on the smooth flow of work, or to explore the effects of changes in the mean times involved. The complete system may be designed on the basis of such explorations.

It should be apparent from this brief discussion of the exploratory use of models that we are making much stronger assumptions about the relationship of the model to the prototype than those necessary for a descriptive use. We are asking for a correspondence of behaviour outside the range over which we have actual empirical evidence.

It should be noted that models sometimes assist in the exploration of situations by their deficiencies as well as their successes. A good example is the data on the growth of the Scottish National Party given in Figure 10.5. The logistic model fits this data very well for the first eight elections. However, on the ninth election the data point fell well below the line. Thus there was a clear indication of a change in the political environment between these two elections, but also of a stability in that environment for the previous eight. Thus the quality of fit provides useful information in exploring the underlying situation.

We now examine the influence of exploratory applications on the processes of statistical modelling.

(a) *Identification*: To justify using a model outside the circumstances in which the empirical evidence was obtained requires that it has strong conceptual as well as empirical support. So there has to be serious thought given to conceptual identification as well as to empirical identification. In descriptive applications it was sufficient to concentrate on what was observed to happen, for exploration the 'why and how' become important. A relationship between two variables that is not correctly identified may cause the model to behave completely differently from the prototype in circumstances not covered by the empirical evidence. The identification process thus needs to include the study of the fine structure of the prototype as well as its general behaviour. For a descriptive application it is sufficient to be able to say what values any exogenous or regressor variables have taken. Within the exploration context there is the need to assess how such variables might behave. This may be done by just putting limits on them, upper and lower reasonable values, or it might require further modelling of these variables. Such further modelling might for example use a simple autoregressive model or a linear trend in time as the model for an exogenous variable. So the identification process requires the creation of either one complex model or a set of simpler linked models.

(b) *Estimation*: In estimation for exploration it is reasonable to seek unbiased estimators with small variance. There are considerable merits in using Bayesian methods since (i) the availability of posterior distributions enables a broader exploration of the model's behaviour, and (ii) the effects of the underlying assumptions can be investigated via a study of different prior distributions.

(c) *Validation*: In the validation process sensitivity analysis becomes particularly important. With a complex model we need to know which parts are most trustworthy, which estimates are most precise, which assumptions most firm. With this done the exploration of the model behaviour can make allowances for the evidence obtained. For example if an estimator is known to have a large variance then a range of different values will need to be assigned to the parameter as part of the exploration process. The amount of effort put into validation needs to be a function of how far the exploration of model behaviour goes beyond the purely descriptive evidence. If, for example, we are considering values of regressor variables or parameters that are only slightly different from those appropriate for the descriptive use of the model, then the descriptive level of validation is probably enough. If, however, we are moving well away from these values, then we need additional evidence to give confidence in the continuing validity of the model for these regions of exploration.

18.4 Prediction

The word 'prediction' is used here to denote the use of a fitted model to predict the values of some variables for given values of others. This definition covers a range of actual model uses. These include:

(a) Pure prediction

Given the fitted linear trend model

$$y_t = \hat{\beta}_0 + \hat{\beta}_1 t + e_t$$

predict y_0 for a given future time t_0.

(b) Smoothing

For this model estimate the expected value, $\beta_0 + \beta_1 t$, at past times, i.e. smooth out the randomness e_t of the past data.

(c) Calibration

This is the problem of estimating the value of a regressor variable given an observation on the dependent variable. For example we may require to estimate the time at which the expectation of y is equal to some value y_0.

In all these cases the essence of prediction is finding values for some, *predicted*, variables, given the values of other, *predictor*, variables. Prediction is one of the most common uses of fitted models and we shall therefore devote some space to illustrating the general methodology before discussing its relationship with the process of modelling.

Example 1 *Prediction with a regression model*

Having fitted a regression model from a set of n observations we shall often use it to predict the value, y_0, that the dependent variable will take if x, the regressor variable, is set at some value x_0.

From the model

$$y_0 = \beta_0 + \beta_1(x_0 - \bar{x}) + e.$$

A natural predictor, $\hat{y}_0$, of y_0 is

$$\hat{y}_0 = \hat{\beta}_0 + \hat{\beta}_1(x_0 - \bar{x}).$$

The quantities $\hat{\beta}_0$, $\hat{\beta}_1$ and e_0 are independent and have expectations β_0, β_1 and zero respectively. So

$$E(\hat{y}_0) = \beta_0 + \beta_1(x_0 - \bar{x})$$

and

$$E(y_0 - \hat{y}_0) = 0.$$

Both y_0 and $\hat{y}_0$ are random variables but it is reasonable to express the last result as a statement that $\hat{y}_0$ is an unbiased predictor of y_0. From previous results we have

$$V(\hat{\beta}_0) = V(\bar{y}) = \frac{\sigma^2}{n},$$

$$V(\hat{\beta}_1) = \frac{\sigma^2}{\sum(x_i - \bar{x})^2}.$$

Hence $\quad V(y_0 - \hat{y}_0) = V[(\beta_0 - \hat{\beta}_0) + (\beta_1 - \hat{\beta}_1)(x_0 - \bar{x}) + e_0]$

$$= \frac{\sigma^2}{n} + \frac{\sigma^2}{\sum(x_i - \bar{x})^2}(x_0 - \bar{x})^2 + \sigma^2 = \sigma_{\tilde{e}}^2,$$

where we denote the prediction error by

$$\tilde{e} = \text{observed} - \text{predicted} = y_0 - \hat{y}_0.$$

If e is normally distributed so will be $\hat{\beta}_0$, $\hat{\beta}_1$, e_0 and hence $\tilde{e}$ will be $N(0, \sigma_{\tilde{e}}^2)$. We may use this fact to construct a 95% confidence interval, called a *prediction interval* for y_0. Estimating σ by the residual mean square, s^2, $\tilde{e}/s_{\tilde{e}}$ will have a t-distribution so the 95% interval will be

$$-t' < \tilde{e}/s_e < t', \text{ where } t' = t_{95\%, n-2}.$$

Substituting for $\tilde{e}$ and $s_{\tilde{e}}$ and rearranging gives

$$\hat{y}_0 - t's\left[1 + \frac{1}{n} + \frac{(x_0 - \bar{x})^2}{\sum(x_i - \bar{x})^2}\right]^{1/2} < y_0 < \hat{y}_0 + t's\left[1 + \frac{1}{n} + \frac{(x_0 - \bar{x})^2}{\sum(x_i - \bar{x})^2}\right]^{\frac{1}{2}}$$

Notice that this interval widens as x_0 moves further from $\bar{x}$.

Example 2 Calibration with a regression model*
To illustrate the calibration problem we shall again use the simple linear model of the Prediction from Regression example. In prediction we take the value of the regressor variable x as given and seek to make statements about the corresponding y. In calibration y is given and interest centres on the corresponding x. A common application of regression models where this arises is in the use of approximate measuring techniques, that have been calibrated against precise techniques. For example, in a preliminary experiment a series of items are measured by a field instrument and by a precision instrument giving observations y_i and x_i respectively. These are used to obtain the fitted calibration model

$$y_i = \hat{\alpha} + \hat{\beta}(x_i - \bar{x}) + \hat{e}_i$$

The field instrument is now used to measure an item and the value y_0 obtained. The question now arises as to how to estimate the corresponding x_0 where

$$y_0 = \alpha + \beta(x_0 - \bar{x}) + e_0.$$

A natural approach would be to use the fitted model, $\hat{x}_0$ representing the estimate of x_0, thus

$$y_0 = \hat{\alpha} + \hat{\beta}(\hat{x}_0 - \bar{x})$$

and hence

$$\hat{x}_0 = \bar{x} + \frac{y_0 - \hat{\alpha}}{\hat{\beta}}.$$

We would obviously like $\hat{x}_0$ to have expectation x_0 and some suitably small variance. However, if we, as usual, assume that the errors are normally distributed, it is found that the expectation of $\hat{x}_0$ cannot be mathematically calculated and its variance is infinite. The mathematics of this will not be examined, but the root of the problem is the occurrence in the formula for $\hat{x}_0$ of $1/\hat{\beta}$ which is the reciprocal of a normally distributed quantity. The distribution of $\hat{\beta}$ includes $\hat{\beta} = 0$ for which $1/\hat{\beta}$ becomes infinite. This has the effect of making the expectation undefined and the variance infinite. There are four ways of dealing with this situation. Experience in the sciences suggests the commonest way is to ignore the problem entirely and simply use $\hat{x}_0$. The second way is to seek alternative estimators that have better properties. These have been sought for at least forty years and there is still no simple universally accepted alternative. A third way is to employ Bayesian methods. These give a clear conceptual treatment but involve theory beyond the scope of this book, see the reference by Aitchison and

Dunsmore. The fourth approach is to seek, not a point estimate, $\hat{x}_0$, but, confidence limits for x_0. This can be done by using the same ideas as those of the prediction interval.

If we define the variable z by

$$z = y_0 - \hat{\alpha} - \hat{\beta}(x_0 - \bar{x}),$$

then, using the results previously used to discuss prediction,

$$E(z) = \alpha + \beta(x_0 - \bar{x}) - \alpha - \beta(x_0 - \bar{x}) = 0,$$

$$V(z) = \sigma^2 + \frac{\sigma^2}{n} + \frac{\sigma^2}{\sum(x_i - \bar{x})^2}(x_0 - \bar{x})^2 = \sigma^2 \left[1 + \frac{1}{n} + \frac{(x_0 - \bar{x})^2}{\sum(x_i - \bar{x})^2} \right]$$

and z is normally distributed.

Estimating σ^2 by the usual residual mean square, s^2, leads to

$$\frac{z}{s\left[1 + \dfrac{1}{n} + \dfrac{(x_0 - \bar{x})^2}{\sum(x_i - \bar{x})^2} \right]^{1/2}} \cap t_{n-2}.$$

Hence putting in 95 % limit values for t leads to a quadratic equation for x_0

$$1 + \frac{1}{n} + \frac{(x_0 - \bar{x})^2}{\sum(x_i - \bar{x})^2} = \left(\frac{z}{st} \right)^2 = \left(\frac{y_0 - \hat{\alpha} - \hat{\beta}(x_0 - \bar{x})}{st} \right)^2.$$

The solutions of these equations, for upper and lower limits of t, provide the values from which the sensible 95 % limits for x_0 can be chosen.

Before leaving this example of calibration, it is worth noting that the word calibration is unfortunately also applied in modelling as synonymous with estimation.

Example 3 *Prediction with an ARMA model**
Consider the time series modelled by

$$x_t - 0.6\, x_{t-1} + 0.2\, x_{t-2} = e_t - 0.5\, e_{t-1}.$$

To predict x_{t+1} we rewrite the model for time $t+1$ so that

$$x_{t+1} = 0.6\, x_t - 0.2\, x_{t-1} + e_{t+1} - 0.5\, e_t.$$

A forecast of x_{t+1}, $\tilde{x}_{t,1}$ is obtained at time t by taking the expected value of x_{t+1}, assuming that at this time we know x_t, x_{t-1} and e_t but that the future e_{t+1} is obviously unknown and its expected value of zero is used. Thus

$$\tilde{x}_{t,1} = 0.6\, x_t - 0.2\, x_{t-1} + 0 - 0.5\, e_t.$$

The *Forecast Error* $e_{f,t,1}$ is

$$e_{f,t,1} = x_{t+1} - \tilde{x}_{t,1} = e_{t+1},$$

on substituting for x_{t+1} and $\tilde{x}_{t,1}$.

Thus the forecast errors for one step ahead forecasts are equal to the random terms e. Hence the term e_t, required for the forecast $\tilde{x}_{t,1}$ can be found as the previous forecast error, $x_t - \tilde{x}_{t-1,1}$.

If we require a forecast, $\tilde{x}_{t,2}$, of x_{t+2} referred to as a forecast with *Lead Time* two, we again use the model:

$$x_{t+2} = 0.6\,x_{t+1} - 0.2\,x_t + e_{t+2} - 0.5\,e_{t+1}$$

The value of x_t is known at time t but using the expected value on the right-hand side gives expected values of zero for both e_{t+2} and e_{t+1} and the forecast $\tilde{x}_{t,1}$ as expected value of x_{t+1}. Hence

$$\tilde{x}_{t,2} = 0.6\,\tilde{x}_{t,1} - 0.2\,x_t$$

The forecast error, $e_{f,t,2}$, will be

$$x_{t+2} - \tilde{x}_{t,2} = (0.6\,x_{t+1} - 0.2\,x_t + e_{t+2} - 0.5\,e_{t+1}) - (0.6\,\tilde{x}_{t,1} - 0.2\,x_t)$$

so

$$e_{f,t,2} = 0.6\,e_{f,t,1} + e_{t+2} - 0.5\,e_{t+1}$$
$$= 0.1\,e_{t+1} + e_{t+2},$$

since

$$e_{f,t,1} = e_{t+1}.$$

Hence
$V(e_{f,t,2}) = 1.01\,\sigma^2$, as e_{t+1} and e_{t+2} are independent.

Notice that these calculations assume that the parameter values $0.6, 0.2, 0.5$ are 'well estimated' constants. To see the effect of recognizing that the parameters are estimated, we examine the simplified model

$$x_{t+1} = \phi x_t + e_{t+1}.$$

The forecast based on a fitted model is

$$\tilde{x}_{t,1} = \hat{\phi} x_t,$$

where $\hat{\phi}$ is the estimated ϕ. Now the forecast error is

$$e_{f,t,1} = x_{t+1} - \tilde{x}_{t,1}$$
$$= \phi x_t + e_{t+1} - \hat{\phi} x_t$$
$$= (\phi - \hat{\phi})x_t + e_{t+1}$$

Thus any error in the estimator $\hat{\phi}$, from the true ϕ, will cause a systematic deviation of the forecast error from e_{t+1}.

Example 4 Prediction with a linear predictor
Suppose that we have a series of observations in time for which we only know, quite accurately from substantial data, that the mean is μ, the variance is σ^2, and the autocovariance of lag one is $\phi\sigma^2$. We can subtract μ from each observation giving a set of data $\{x_t\}$ having zero mean. Suppose we now seek to predict x_{t+1} by a function of the last observation. The

simplest linear function is

$$\hat{x}_{t+1} = ax_t.$$

We must now choose the constant, a, to obtain a good forecast. To do this we need to define what is meant by a good forecast. A fairly natural criterion is that the forecast has a small expected error squared, i.e. $S = E[(x_{t+1} - \hat{x}_{t+1})^2]$ is small. Thus we minimize S. Now

$$S = E[(x_{t+1} - ax_t)^2] = E(x_{t+1}^2) - 2a\, E(x_{t+1}x_t) + a^2 E(x_t^2)$$
$$= \sigma^2 - 2a\,\phi\sigma^2 + a^2\sigma^2.$$

This is minimum when a is equal to ϕ. So the best linear predictor using only x_t is

$$\hat{x}_{t+1} = \phi x_t.$$

Notice that the form of this corresponds to the forecast for the AR(1) model of the previous example, but we have here assumed very little about the underlying structure, just that it has an autocovariance. If we took the empirical version of the criteria given a set of $n+1$ observations the criterion would become

$$\sum_{t=1}^{n} (x_{t+1} - a x_t)^2,$$

which is minimum for

$$\hat{a} = \frac{\sum x_{t+1} x_t}{\sum x_t^2},$$

which is the method of moments estimate of ϕ.

Example 5 *A Bayesian approach to prediction**

This approach to prediction avoids the problems associated with using a model with estimated parameters that are treated as though they were true values. The future observation, x_T, will have a distribution, depending on these parameters, θ, which can be written as $f(x_T|\theta)$. The prior distribution $h(\theta)$ together with previous data, appearing via the likelihood, $\mathscr{L}(\theta|x)$, give the posterior distribution for θ, $h(\theta|x)$. The joint distribution of x_T and θ given the data can be written as

$$f(x_T, \theta|x) = f(x_T|\theta)\, h(\theta|x).$$

If this is summed (or for a continuous distribution integrated), over all values of θ, then we obtain $f(x_T|x)$, the *Predictive Distribution* of x_T given x. From the sequence of the calculation it will be seen that this approach allows for the unknown values of the parameters by using the prior and posterior distributions. It then removes the need to use explicit estimates of the parameters by in effect, averaging over all possible values of θ, since the last stage could be expressed as an expectation with respect to the posterior distribution, denoted by

$$f(x_T|x) = \underset{\theta|x}{E}[f(x_T|\theta)].$$

Consider again the Poisson process, $P(vt)$, discussed in section 15.2, with r events in time t. It was shown that with a Gamma Prior, corresponding to R events in time T, the posterior distribution of the mean rate of occurrence of events was also a Gamma, corresponding to a total time $T*$ $= T + t$ containing $R* = R + r$ events. If we have a future time interval t' the number of events, r', will have a Poisson distribution, $P(vt')$. Combining this with the Gamma posterior distribution of Chapter 15 leads, after some calculation, to a predictive distribution for r' that has a *Negative Binomial* form

$$Pr(r') = \binom{r' + R* - 1}{r'} p'^{r'} (1-p)^{R*},$$

where $p = t'/(T* + t')$. Figure 18.1 illustrates the situation. We showed in section 15.2 that the Prior evidence, R events in time T, and the empirical evidence, r events in time t, combined to give a posterior distribution equivalent to $R* = R + r$ events in time $T* = T + t$. We now consider the posterior time interval $T*$ linked to the future interval t' giving $T* + t'$. Imagine an experiment involving observing both intervals, posterior and predicted. The constant p is the probability of a random event in the time interval $T* + t'$, (AC), actually falling in the future interval t', (BC). We thus visualize the situation as two adjacent boxes, (AB), (BC), into which we throw 'events', with p as the probability of the event landing in (BC), $1 - p$ the probability of it landing in (AB). We keep throwing until the known $R*$ events have landed in (AB), r' is then the number in (BC). We can use the Binomial formula to give the probability of r', with the reduction of $n = R* + r'$, in $\binom{n}{r}$, by 1 since the last event is known to be in (AB). This distribution is the *Negative Binomial*. The mean of this predictive distribution is

$$R* \frac{(1-p)}{p} = R* \cdot \frac{t'}{T*}$$

which gives an intuitively sensible single value prediction for r'. The book by Aitchison and Dunsmore discusses these methods and their application to both prediction and calibration.

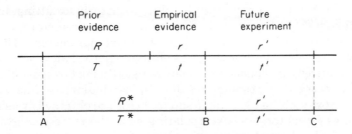

Figure 18.1 The interpretation of a Bayesian predictive distribution

Let us now use the previous examples to illustrate the appropriate processes of statistical modelling.

(a) *Identification*: It is apparent that prediction is a much more limited exercise than either description or exploration. The model does not need to behave generally like the prototype, there is no need to have analogous elements or structures between model and prototype. All that is required is that predictors derived from the model are close to the actual values given by the prototype. Many of the most successful methods of prediction are based on the use of simple empirical models, linear trends or ARMA models. There is no attempt at a conceptual approach. A simple model is fitted as well as possible to recent data and extrapolated to predict values in the near future. Such direct extrapolation of the past into the future is called *Unconditional Prediction*. The methods of identification for short-term forecasting are the simplest needed to obtain the main structure in the recent data, assuming that this structure is unlikely to change much over a short period of time. For very long lead times identification will involve looking for mathematical shapes such as growth curves in the data and in data from analogous situations. It is unlikely that a complex model assuming many internal relationships will predict very well, since such relationships tend to change with time. For a medium lead time knowledge of such relationships might prove useful and a more structured model might give good forecasts.

It is evident that the identification process will not only be influenced by the predictive nature of an application but also, for prediction in time, the distance into the future being predicted.

(b) *Estimation*: The quality of a prediction, x, is usually represented by the distribution of the prediction error, $e_t = x_t - \hat{x}_t$. If we wish to turn this into single measures we might seek unbiased predictions, with $E(e_t) = 0$, for which the variability is as small as possible. Thus we might require a small mean square error, $E(e_t^2)$, or possibly a small mean absolute error, $E(|e_t|)$. A further sensible requirement is that the errors form a series of independent variables, for if they did not we could model this series, forecast future errors and construct improved forecasts. We thus see that prediction problems generate their own criteria that do not immediately relate to criteria for fitting. Indeed in the example on linear prediction we derived a predictor $\hat{a} x_t$ without any reference to the estimation of model parameters. There are, however, some theoretical results which suggest that, where estimates are required, the common estimators usually lead to sensible, though not necessarily optimum, results. In particular for linear models predictors with smallest mean square error are obtained by using least squares estimators and extrapolating as in the regression example. It can also be shown that the procedure of setting future errors at their zero expected values, as done in the ARMA example, also gives minimum mean square error predictors, when the parameters are assumed known.

(c) *Validation*: In a sense validation is irrelevant for prediction models. We are not really interested in whether the model corresponds in any sense to the prototype. Our sole concern is whether $\hat{x}_t$ will be close to x_t. Having said that, it is appropriate to consider whether the assumptions and structure of the prediction model are such that it is likely to continue to give reasonable predictions. These questions revolve essentially round the stability of the prototype. In practice it is wise to monitor this stability by setting up validation procedures for the working data as discussed in section 17.6. The various tracking signals discussed there can be applied directly to the prediction errors.

Before leaving the subject of prediction it is advisable to make one general point. We have discussed the use of models to obtain numerical predictors. In such situations the model is developed for this purpose. A quite different aspect of prediction is the point often made in the literature that for *any* model a test of its quality is its predictive ability. This is reasonable as a general statement, obviously if a model completely fails to predict what is observed qualitatively and quantitatively then we are justified in being very dubious about the model. However, in modelling complex situations a model may be very helpful in giving insights into general forms of behaviour but, because of its complexity and the number of sources of error, it may not be as good predictively as a much simpler model designed for prediction. Thus ability to predict quantitatively is not an overriding consideration for *all* models.

18.5 Decision

Our final class of applications is that which relates to decision making. Such applications may arise as an application of a standard model or the whole model may be developed as a decision-making tool. We first illustrate these two types of application. In the former situation the model is often used to find the value of a variable that is optimum in some sense, or that corresponds to some quantity of interest.

Example 1
Observations are made on the output, y, of a chemical process for different amounts, x, of an added catalyst. A quadratic model

$$y = \alpha + \beta x + \gamma x^2 + e$$

is fitted. The decision is to choose the value x^* of x that maximizes the output. Differentiating y with respect to x and equating to zero, to give the maximum, gives

$$x^* = -\frac{1}{2}\frac{\beta}{\gamma}.$$

The maximum output is then, by substitution,

$$y^* = \alpha - \frac{\beta^2}{4\gamma}.$$

Given the fitted model with estimates $\hat{\alpha}$, $\hat{\beta}$ and $\hat{\gamma}$ these values would become

$$\hat{x}^* = -\frac{1}{2}\frac{\hat{\beta}}{\hat{\gamma}}, \quad \hat{y}^* = \hat{\alpha} - \frac{\hat{\beta}^2}{4\hat{\gamma}}.$$

In the second type of decision application we consider the modelling of a problem involving choice between alternative decisions. To make such decisions we need some measure of the value, to us, of each of the possible results of the decisions. In the previous example this could be expressed negatively as the loss in output below the true optimum. Sometimes it may be expressed as a positive profit in money terms. On other occasions it may be some intuitive quantification of more nebulous value, such as quality of the environment. Such a measure is generally referred to as the *Utility* of an outcome.

Example 2*
Consider the case where the underlying situation can be in only one of two states S_1 and S_2; for example, an oil rig is either sited over oil or it is not. Faced with this we only make one of two decisions, d_1 or d_2, to drill or not to drill. Usually we denote d_1 as the sensible decision if S_1 is true and similarly, d_2 for S_2.

We assume that we possess information on potential costs and profits and can thus construct what is often called a *utility table*.

Utility		Decision	
		d_1	d_2
State	S_1	u_{11}	u_{12}
	S_2	u_{21}	u_{22}

In this table u_{ij} is the utility of making decision d_j when S_i is the true state. Let us assume that we can also use past experience to assign prior probabilities $p(S_1)$ and $p(S_2)$ $(p(S_1) + p(S_2) = 1)$ to the two states S_1 and S_2 respectively. If some relevant data, D, is available we can use it to move from prior probabilities to posterior probabilities, $p(S_i|D)$, using the methods discussed in section 12.4. It was shown there that

$$p(S_i|D) \propto \mathscr{L}(S_i)p(S_i)$$

where $\mathscr{L}(S_i)$ is the likelihood of the state S_i which, remember, is also the probability, or probability density of the data assuming S_i to be true. In the oil example D might refer to the outcome of some geological test for which we could evaluate $Prob(D|S_i)$ and hence, equivalently $\mathscr{L}(S_i)$. The basic rule for reaching decisions, called *Bayes Rule*, is to choose the decision, d_j, that leads to the largest expected utility. The expected utility, if d_1 is chosen, is

$$E(\text{Utility}|d_1) = u_{11}p(S_1|D) + u_{21}p(S_2|D).$$

If d_2 is chosen it is

$$E(\text{Utility}|d_2) = u_{12}p(S_1|D) + u_{22}p(S_2|D).$$

Thus decision d_1 will be chosen if

$$u_{11}p(S_1|D) + u_{21}p(S_2|D) > u_{12}p(S_1|D) + u_{22}p(S_2|D)$$

i.e. if

$$\frac{u_{11}-u_{12}}{u_{22}-u_{21}} \cdot \frac{p(S_1|D)}{p(S_2|D)} > 1.$$

Relating back to the prior probability and likelihood gives the *Decision Rule*: choose d_1 if

$$\frac{u_{11}-u_{12}}{u_{22}-u_{21}} \cdot \frac{\mathscr{L}(S_1)}{\mathscr{L}(S_2)} \cdot \frac{p(S_1)}{p(S_2)} > 1.$$

Clearly d_2 is chosen if this ratio is less than one.

It can be seen that this ratio depends on three elements

(i) $\dfrac{u_{11}-u_{12}}{u_{22}-u_{21}} = \dfrac{\text{value of correct choice if } S_1 \text{ is true}}{\text{value of correct choice if } S_2 \text{ is true}}$

Thus, other things being equal, we would make the decision d_1, corresponding to S_1, if the difference in utility between right and wrong choices was greater if S_1 were true than if S_2 were true. We chose the decision that is best for the state where the relative cost of a wrong decision is greatest.

(ii) $\dfrac{\mathscr{L}(S_1)}{\mathscr{L}(S_2)} = $ the likelihood ratio

Thus, other things being equal, we would make the decision corresponding to the most likely state.

(iii) $\dfrac{P(S_1)}{P(S_2)} = $ the prior odds.

Thus, other things being equal, we would make the decision corresponding to the highest prior probability.

In practice we would combine all three ratios as the basis for a decision. Notice that the decision rule is simply the Bayes rule, considered in subsection 12.6.1, multiplied by a term, (i), that allows for utilities. So the rule can be applied to the choice of hypotheses or of models in situations where we have information available on costs.

Having briefly illustrated the decision-making application of models, we turn to the effect of such a role on the process of modelling.

(a) *Identification.* An immediate consideration here is that we may have to model not only the prototype but also the utility function. This may be just a matter of assigning values, as in the last example, or it might, in a decision problem involving forecasting, involve finding a function to represent the financial cost of forecast errors. The level of detail required and the balance of empirical and conceptual approaches may well only become evident during the iteration of a modelling exercise. Sometimes an analysis leads to a decision so clear and convincing that virtually all reasonable models must give the same decision. If, however, the choice of the best decision is not clear, more detailed identification will be required.

(b) *Estimation*: Both the examples discussed emphasize that the application of standard point estimation methods are not necessarily helpful in decision making. In the first example the natural estimate of the optimum value was $\hat{x}^* = \hat{\beta}/2\hat{\gamma}$, where $\hat{\beta}$ and $\hat{\gamma}$ would usually be least squares estimators. If the regression error is normally distributed then x^* is the ratio of two normals. We noted in the discussion of calibration that this ratio can be a very difficult quantity to handle. Further, using least squares in this problem ignores the fact that the problem has its own natural criterion, in the loss of output due to not choosing the true α and β. If $\tilde{x}$ is an estimate of the optimum x then the actual expected output is

$$\tilde{y} = \alpha + \beta\tilde{x} + \gamma\tilde{x}^2.$$

The loss in expected output is thus

$$l = y^* - \tilde{y}.$$

If we write $\tilde{x} = x^* + d$ then a little algebra will show that $l = d^2$. If we used estimators $\hat{\alpha}$, $\hat{\beta}$, $\hat{\gamma}$ to obtain $\tilde{x} = \hat{\beta}/2\hat{\gamma}$ then

$$d = -\frac{1}{2}\frac{\beta}{\gamma} + \frac{1}{2}\frac{\hat{\beta}}{\hat{\gamma}}.$$

Thus the loss criterion appropriate to this problem is

$$l = \frac{\gamma}{4}\left(\frac{\beta}{\gamma} - \frac{\hat{\beta}}{\hat{\gamma}}\right)^2.$$

Hence using least squares estimators does not here lead to a minimum of the expected loss. In the Bayesian decision example there is no use of point estimators. The data is used naturally via the posterior distribution.

(c) *Validation*: As with identification, the validation requirements will relate closely to the specific decision being made. So in the first example the constant term α played no direct part in the decision. Thus the validation of the fitted model would concentrate on the role of the terms in x and x^2. In the second example examination of the validity of the prior probabilities and of the utilities would need to take place alongside that of the likelihoods.

18.6 The dangers of separating modelling and application

It should be clear from the previous sections of this chapter that the purpose for which a model is constructed has significant impact on the process of modelling. Having said this, it has to be admitted that insufficient consideration is given in current practice. For the most part a model is constructed using routine methods of identification and estimation and then used for any sort of application as though it were a true description of the prototype. Such an approach can lead to a number of dangers and difficulties.

(a) The form of the model may be inappropriate for the particular application.

   (i) Different applications require models working at different levels of detail and aggregation. A model to be used for a weekly decision on stock levels, for a wide range of items, will need to be simple and robust in form and to be based on weekly data. A more sophisticated model that includes regressor variables and uses monthly data just does not do the job required.

  (ii) The form of model may involve implicit assumptions that are inappropriate for the use. For example, decisions based on the use of the model may affect new data in such a way that an assumed independence of successive error terms no longer holds.

 (iii) A sophisticated model may require to be refitted for a particular application, but data of a quality and quantity required may not exist for the new situation. This difficulty occurs particularly where models are originally devised during major research projects, when there are significant funds for data collection, and then later are proposed for use in smaller problem-solving exercises.

 (iv) Where models are to be used by people lacking in statistical expertise, it is important that the form, variables and parameters of the model are intelligible and meaningful to the user. A model that is just a magic formula to the user may give good results at first but, without intelligibility, it will easily be misapplied.

(v) Models developed under the highly controlled conditions of a laboratory or field experiment may not be valid in a less controlled practical situation. This can result in important variables in the practical application being ignored. Conversely a variable whose effect was well estimated in the controlled situation may vary very little in the data of the application, with a consequent difficulty in estimating its effect and in fitting the complete model.

(vi) Though the model may give a well-fitting description of the unconstrained situation, its practical use may lead to changes in that situation. A forecasting model may lead to such bad forecasts that management works hard to ensure that the situation is changed so that the forecasts are not fulfilled. So the application requires that the model be modified, to allow for such events, in its form and its fitting.

(b) The previously estimated parameters of a fitted model may be inappropriate for a given application.

(i) The original method of estimating the parameters might for example be least squares, but the application might involve elements suggesting a non-symmetrical function of the residuals as a criterion of fit. A problem here is that sometimes fitted models are quoted without any indication of the fitting method used or the standard errors of the estimates. Without this information the user is well advised to refit the model.

(ii) Often several different models are brought together as part of a larger model. The overall quality of the larger model does not necessarily come by using 'best fitting' parts, since the interactions between parts have been ignored in the previous estimation.

(c) The final danger in separating modelling from application is a psychological one. The model is taken as given, handed down from on high. It can easily come to be treated as though it represents the whole Truth about the prototype. The dangerous corollary of this is that the real situation is treated as though it behaved like the model. The social sciences have generated many models of which too much was expected. When they failed to live up to expectation they were then discarded as futile. A helpful counter to these problems is to consider not the model but a number of alternative models, of perhaps varying quality, but operating at different levels of detail, and perhaps based on different conceptual bases. The fact that one is selected for a particular application should not alter the awareness that there were other possible models for describing the prototype.

Part VII

Iteration

Chapter 19

Developing models

19.1 Introduction

As we indicated in Chapter 1 modelling is not a process that we can carry through in a formal ordered fashion. It is a highly iterative activity with frequent second thoughts and new discoveries that take us back to rethink previous stages. Having now covered the basic ground of models, identification, estimation, validation, and application, we devote the final part of this book to consider ways in which we might iterate these processes and improve our modelling. In this chapter we shall look in general terms how the list of well understood and documented models has developed over the years, for iteration is a historical process as well as one carried out by individual modellers. In the following chapters we illustrate in particular how the processes of identification and estimation can be developed once we have an initial fitted model.

19.2 Distributional models

In the historical development of the models, that form part of the tool kit of a modeller, a simple model is first developed and, if successful, it is repeatedly modified and generalized to deal with new situations that arise. To illustrate this idea we examine a series of examples related to (a) the Poisson Distribution and (b) some specific Regression and Progression Models. We shall explore these examples in terms of identification and estimation as well as in terms of basic model structure.

Developing the Poisson Distribution

Example 1 *The initial model.*
Historically the first set of data to which a Poisson model was fitted was that shown in Table 19.1. Baron von Bortkewitch collected data on the number of cavalrymen kicked and killed by army horses in each of a

286

Number of deaths per year per corps	0	1	2	3	4	5+
Frequency	109	65	22	3	1	0
Expected frequency	109	66	20	4	1	0

Table 19.1

number of years in each army corps. He suggested the Poisson model as a suitable formula. The expected values of the fitted Poisson frequencies and the original data are given in Table 19.1.

Example 2 Mixing models
The data below gives the number of blemishes, caused by a blight, per plant on 217 plants.

Number	0	1	2	3	4	5	6 or more
Frequency	150	39	17	8	2	1	0

$n = 217$

A graphical study or a goodness of fit test with this data raise doubts about it being Poisson. An indication of the nature of the divergence from the Poisson can be obtained by making use of the fact that for a Poisson

$$(r+1)\frac{Pr(r+1)}{Pr(r)} = \mu.$$

Replacing $Pr(r)$ by an estimate, the frequency $f_r/217$, we can tabulate as follows

$$r = 0 \quad 1 \quad 2 \quad 3 \quad 4 \quad 5$$

$$\frac{(r+1)f_{r+1}}{f_r} = \frac{39}{150} \quad \frac{34}{39} \quad \frac{24}{17} \quad \frac{8}{8} \quad \frac{5}{2} \quad 0.$$

It is clear from this that, allowing for random fluctuations, the ratios for $r = 1$ to 5 look reasonably steady but that the frequency f_0 is far higher than one would expect for the Poisson with a parameter μ close to one. A conceptual modelling approach to this problem suggests that an explanation might be that not all the plants are susceptible to the blight. Suppose that a proportion p are not susceptible, then the group of plants showing no blemishes contains all the non-susceptibles together with the susceptible plants for which the Poisson gives $r = 0$. Hence

$$Pr(R = 0) = p + (1-p)e^{-\mu}$$

similarly

$$Pr(R = r) = (1-p)e^{-\mu}\frac{\mu^r}{r!} \quad r = 1, 2, \ldots.$$

Thus the population under consideration is a mixture of two different populations. The distribution of R is an example of a Mixed Distribution. To estimate the two parameters by maximum likelihood requires numerical iteration. As a first approximation we can use methods based on analogy. For $r > 0$ the ratios of frequencies for the Poisson still holds so an estimate of μ using the frequencies for $r = 1$ and $r = 2$ is, from the previous table, $\hat{\mu} = 34/39$. The proportion of observations in the zero frequency gives the second equation, since we can write

$$\hat{p} + (1-\hat{p})e^{-\hat{\mu}} = \frac{150}{217},$$

which on solving gives $\hat{p} = 0.47$ as a first approximation.

Example 3 A Compound Distribution
Many practical problems involve relating the Poisson model to other distributional models. Suppose for example that seeds land at random over a certain area A, the number of seeds is thus $P(vA)$ where v is the mean seed density. If θ is the probability of a seed germinating and growing, then we can use the Binomial model to say that the number of plants, s, will be $\text{Bin}(r, \theta)$, where r is the actual number of seeds in the area A. So we have a Binomial distribution in which the number of trials is itself a random variable from a Poisson. The probability of s plants is given, using the 'chain rule' of Appendix 1, by

$$Pr(s) = \sum_{r \geqslant s} Pr(s|r)\, Pr(r),$$

where $Pr(s|r)$ is the Binomial Probability conditional on r and $Pr(r)$ is the Poisson Probability. A certain amount of algebra will show that this is in fact a Poisson distribution with mean θvA. This is intuitively right. Imagine the initial cloud of seeds falling at random. The Binomial part of the model in effect reduces the density of this cloud, in terms of plants produced, by a factor of θ, but we will still have a Poisson form of random occurrences. Distributions which arise when the parameter of one distribution is the random variable of another are called *Compound Distributions*. Notice that such distributions are, in Bayesian terms, the Posterior Distributions in cases where the Prior Distribution has a clear interpretation in terms of the prototype.

Example 4 Generalizing
It will be remembered that when the Poisson Distribution was introduced in section 3.3 it was shown as arising by counting the number of events, R, in a Poisson Process, within fixed intervals. Often the events $E_1, \ldots, E_R$ have random variables, $X_i \ldots, X_R$, associated with them.

(i) The event might be the dropping of a seed on a unit area of ground. The seed may germinate and grow into a plant, indicated by $x_i = 1$, or it may not, indicated by $X_i = 0$. If R is the number of seeds dropping at random then R will be a Poisson random variable and the number of germinated seeds, N, will be

$$N = \sum_{i=1}^{R} X_i.$$

(Note the relation of this illustration to the alternative presentation of Example 3.)

(ii) Instead of individual seeds we might be considering seed pods each pod containing X_i seeds, where X_i is itself Poisson distributed. The total number of seeds in a unit area will thus be

$$N = \sum_{i=1}^{R} X_i.$$

(iii) If the Poisson process represents the arrival of insurance claims of a certain type and R is the number arriving in a month. Each claim will be for an amount X_i. The total amount, N, claimed in a month will be

$$N = \sum_{i=1}^{R} X_i.$$

The distribution of N is referred to as a *Generalized Poisson*. Clearly the actual distribution of N will depend on the form of the distribution of X.

Example 5 Truncating

The survey results in Table 19.2 show a discrete distribution with a fairly rapid decrease in probability. This at first sight might be a Poisson model with a fairly low mean. However, in a Poisson model the variable r, the number in a household, can take any value 0, 1, 2, ... In this case the value $r = 0$ is not a possibility. This suggests that we modify the Poisson model to exclude $r = 0$. This may be done by making use of the rules of conditional probability as we are interested in $Pr(R = r | r \neq 0)$. Thus

$$Pr(R = r | r \neq 0) = \frac{Pr(R = r \text{ and } r \neq 0)}{Pr(r \neq 0)}.$$

Table 19.2

Distribution of household size from a survey of households receiving a particular social service

Number of people in household	Number of households
1	436
2	133
3	19
4	2
5	1
6	0
7	1
8+	0

Starting from the Poisson

$$Pr(R = r) = e^{-\mu} \frac{\mu^r}{r!},$$

the probabilities for $r = 1, 2, \ldots$ in $Pr(R = r$ and $r \neq 0)$ are given by this expression. Further

$$Pr(r \neq 0) = 1 - Pr(R = 0) = 1 - e^{-\mu}.$$

Thus if we start with a Poisson and then impose the condition $r \neq 0$ we obtain the *Truncated Poisson*

$$Pr(R = r) = \frac{e^{-\mu} \dfrac{\mu^r}{r!}}{1 - e^{-\mu}} \qquad r = 1, 2, \ldots.$$

The shape of this distribution is identical to the Poisson for $r = 1, 2, 3, \ldots$, the divisor $(1 - e^{-\mu})$ is simply a constant that ensures that

$$\sum_{r=1}^{\infty} Pr(R = r) = 1.$$

This divisor also has the effect of increasing the expectation of the Poisson to

$$E(r) = \frac{\mu}{1 - e^{-\mu}}$$

To understand the significance of this model, remember that the Poisson is a model of purely random occurrences. We can perhaps visualize the model as follows, using a value of $\mu = 0.6$ (which we will justify later). Imagine a million houses and a population of 600,000 people. Each person is allocated at random to one of the houses, independently of all other people's allocation. There will thus be a mean of 0.6 persons per house. A proportion, $e^{-0.6}$ which is about 55%, of the houses will be unoccupied. These unoccupied houses are now removed from the population and a sample of 592 of the remaining occupied houses is now chosen at random to give our data.

This form of modification of a distribution in which part of the distribution is cut off is called *Truncation*. In the example here we have truncated at the lower end of the distribution. In situations where for example vehicles arrive at random at a car park or telephone calls arrive at a switchboard without a queueing facility, the numbers of cars or calls is Poisson with truncation in the right-hand tail, at the maximum number of cars or calls that can be dealt with.

Example 6 Censoring
Certain types of fast counters for atomic particles operate in a binary fashion. Thus, if in a millisecond period zero particles are detected (00) is

recorded, one particle records (01), two particles record a (10). However, (11) indicates that three *or more* particles have been detected. In a one-second period the counter will indicate that in $f_i (i = 0, 1, 2)$ millisecond periods i particles were recorded and that in f_c periods three or more particles were detected. We thus have a situation similar to that of truncation except that we do know f_c, the total number of observations in the truncated, censored, part of the distribution. Assuming particles arrive as a Poisson process we have probabilities

$$Pr(R = 0) = e^{-\mu}, \; Pr(R = 1) = e^{-\mu}\mu, \; Pr(R = 2) = e^{-\mu}\frac{\mu^2}{2!}$$

$$Pr(R = 3 \text{ or more}) = 1 - Pr(R = 0, 1 \text{ or } 2) = 1 - e^{-\mu}\left(1 + \mu + \frac{\mu^2}{2}\right).$$

Thus we can write the likelihood as

$$\mathscr{L}(\mu) = e^{-f_0\mu} \cdot e^{-f_1\mu} \cdot e^{-f_2\mu}\frac{\mu^{f_1 + 2f_2}}{2^{f_2}} \cdot \left[1 - e^{-\mu}\left(1 + \mu + \frac{\mu^2}{2}\right)\right]^{f_c}$$

$$\propto e^{-n\mu}\mu^{f_1}\mu^{2f_2}\left[e^{\mu} - \left(1 + \mu + \frac{\mu^2}{2}\right)\right]^{f_c}$$

where $n = f_0 + f_1 + f_2 + f_c$.

The form of this requires numerical solution to obtain an estimate of μ. However, the example shows that censoring does not give rise to major problems but leads to a fairly straightforward development of the basic distributional models.

19.3 Regression and progression models*

In the same way as the family of distributional models has developed so has the family of regression and progression models. In the field of regression some of the major recent developments concern the use of regression models with non-normal dependent variables, examples of which were dealt with in section 6.3. There has been parallel development of other common models in which various of the usual assumptions are dropped. There has also been steady development of mixed models, for example regression and simultaneous equation models with errors from an ARMA process. We shall focus in this section on just one, rather different, aspect of the development of models. This aspect is essentially the application of the 'principle of alternatives' of section 9.7. There have been very rapid developments when old models have been viewed in new ways, particularly new ways that lend themselves to modelling on a computer. The following examples provide some elementary illustrations of such reformulation.

Example 1

In Table 9.1 we devised a model for an electrical circuit in the form

$$LC\frac{d^2v_o}{dt^2} + RC\frac{dv_o}{dt} + v_o = v_i$$

where v_i and v_o were the input and output voltages respectively. Thus the system is described in terms of the rate of change, $\frac{dv_o}{dt}$, and $\frac{d^2v_o}{dt^2}$. An alternative view of this system is to use two variables to describe it. We have already noted that the current relates to the rate of change of voltage so

$$\frac{dv_o}{dt} = \frac{1}{C}i. \qquad (a)$$

hence

$$\frac{d^2v_o}{dt^2} = \frac{1}{C}\frac{di}{dt}.$$

The above system equation can thus be rewritten as

$$\frac{di}{dt} = -\frac{R}{L}i - \frac{1}{L}v_o + \frac{1}{L}v_i \qquad (b)$$

Thus if we use the two variables, v and i, we can describe the systems behaviour in terms of the two equations (a) and (b) that give simply the rates of change of v_o and i in terms of linear expressions in v_o, v_i and i. The variables v_o and i are called the *State-Variables* for the system. The form of the equations show how the state of the system, described by these variables, changes with time. These equations are called the *System Equations*. Introducing some statistical reality into the situation we might introduce new variables x and y representing the observed values of v_o and i. If we allow only for simple additive errors e_v and e_i respectively then we obtain two *Observation Equations*

$$x = v_o + e_v$$
$$y = i + e_i.$$

Example 2

In section 9.8 we briefly mentioned that the linear model

$$x_t = \alpha + \beta t + e_t$$

could be reformulated using μ_t to represent $E(x_t)$ and writing

$$\mu_t = \mu_{t-1} + \beta.$$

This can be seen, in the terminology of the last example, as state equations

$$\mu_t = \mu_{t-1} + \beta,$$
$$\beta = \text{constant},$$

and observational equation

$$x_t = \mu_t + e_t.$$

The state equations are purely deterministic. A natural generalization of this model is to allow random variation in the states defined by μ and β.

Thus we can write

$$\mu_t = \mu_{t-1} + \beta_t + u_t$$
$$\beta_t = \beta_{t-1} + v_t$$

where μ_t and β_t are the mean and slope at time t and u_t and v_t are zero mean innovations. The model thus represents a local linear trend in which both mean and slope are subject to random perturbations.

Chapter 20

Improving identification

20.1 Introduction

The easiest way to examine the curvature of a line is to compare it with the edge of a ruler. Similarly the best way to examine the complexities of a set of data is to use a simple fitted model as a basis for comparison and investigate where the data fails to fit. It is rarely worth while to go directly to a sophisticated and elaborate model. So our initial models will usually contain assumptions such as linearity, homogeneity, independence, and normality. Comparison with data will often show that these assumptions are wrong. We thus need to remodel. Such remodelling is an example of the general process of iteration. As we have indicated before modelling is rarely a process that we carry through in a formal ordered fashion. It is a highly exploratory activity with frequent second thoughts and new discoveries that cause us to rethink a previous stage. Having covered the basic ground of identification, in Part III we shall devote this chapter to consider ways in which we might iterate and reconsider this process.

Many of the techniques discussed in this chapter will be of immediate applicability when carrying out eclectic identification or validation studies. For example in the next section we suggest the construction of models for sets of residuals as a way of improving the initial identification. We introduced residual analysis as a means of checking the initial model. We shall see now that, if the residuals do not show the randomness that would indicate a satisfactory model, their pattern might actually suggest ways of improving that model. Any approach to checking models that gives such helpful diagnostics is referred to as *Diagnostic Checking*.

20.2 Residual modelling

In many of the models considered we have finished with equations of the form

$$y_i = g(x; \hat{\theta}) + \hat{e}_i,$$

293

where y_i is an observation, $g(x; \hat{\theta})$ is the fitted model, based on data x and estimated parameters $\hat{\theta}$, and $\hat{e}_i$ is the residual, or possibly forecast error. As we have repeatedly emphasized, any error in the choice of the model or its fitting to the data will reveal itself in the residuals. To take a simple example suppose the 'true' situation is

$$y = 23 + 24x + 3x^2 + e,$$

and a fitted linear trend is obtained

$$\hat{y} = 22 + 25x.$$

We thus have residuals

$$\hat{e}_1 = 1 - x_i + 3x_i^2 + e_i.$$

As our estimation procedure was effectively out by $1 - x_i$ it is unlikely that we would be able to detect this term against the randomness introduced by the e_i. However, the $3x^2$ term is due to a wrong selection of model and should show if we plot the residuals against x_i. Such a plot could be made clearer if we try to reduce the effect of the errors e_i. This may be done by a smoothing technique. One simple way of doing this is to take a moving section of 3 or 5 observations working along the x axis picking up one and dropping one observation at each step. At the middle point of the section a representative value of $\hat{e}$ is plotted. This could be the average value of the residuals in the section, in which case the technique is referred to as using moving averages. Alternatively and more simply the median value could be used. Such smoothing will help to reveal any remaining structure in the residuals. Suppose we detect a structure that is apparently quadratic. Because of the nature of the situation we have already allowed for constant and $\beta_1 x$ terms so we can concentrate on the quadratic alone. Thus a model for the residuals would be

$$\hat{e}_i = \beta_2 x_2^2 + f_i$$

where f_i is the independent random error for this new model. The least squares estimate of β_2 is, from our previous work,

$$\hat{\beta}_2 = \frac{\sum \hat{e}_i x_i^2}{\sum x_i^4}.$$

This could now be checked for its statistical significance, against the hypothesis $\beta_2 = 0$, and its technical significance in the situation being modelled. Suppose we found a significant $\hat{\beta}_2$ of $\hat{\beta}_2 = 2$, we then have two models

$$y_i = 22 + 25x_i + \hat{e}_i,$$
$$\hat{e}_i = 2x_i^2 + \hat{f}_i.$$

Combining gives

$$y_i = 22 + 25x_i + 2x_i^2 + \hat{f}_i.$$

We have thus built a second model for y by modelling the residuals using a simple regression model.

Another situation that occurs frequently is that, though no deterministic structure is found in the residuals, some form of stochastic structure is found. Consider two simple possibilities. The residuals might show a simple moving average structure so that the residual model is

$$\hat{e}_i = f_i - \theta f_{i-1}.$$

Hence if we have the linear model considered previously as the initial model and if the estimate of θ was $\hat{\theta} = 0.4$ we would have the combined fitted model

$$y_i = 22 + 25x_i + f_i - 0.4f_{i-1}.$$

Alternatively the residuals might show an autoregressive structure

$$\hat{e}_i = \phi\hat{e}_{i-1} + f_i.$$

If in this case we estimated ϕ by $\hat{\phi} = 0.4$ the second fitted model is

$$\hat{e}_i = 0.4\,\hat{e}_{i-1} + f_i.$$

To combine this with the initial model we need to first re-organize to get a term $\hat{e}_i - 0.4\hat{e}_{i-1}$. It will be seen that finding $y_i - 0.4y_{i-1}$ will achieve this, giving as the single model

$$y_i = 0.4\,y_{i-1} + 13.2 + 25x_i - 10x_{i-1} + f_i.$$

Yet another possibility is that the residuals show a systematic change in variance as x changes. A plot of $|\hat{e}_i|$ against x_i will reveal this. As an example suppose this plot shows a linear increase so that we might model the standard deviation of $\hat{e}_i$ by

$$\sigma(\hat{e}_i) = x_i\sigma.$$

It follows that the quantity $e_i/x_i = f_i$, satisfies the usual requirement of homogeneity, i.e. $V(f_i) = \sigma^2$. Again we wish to combine models and this is achieved by dividing through by x_i to give

$$\frac{y_i}{x_i} = \frac{22}{x_i} + 25 + f_i.$$

This combined fitted model gives the independent errors of constant variance in their usual position in the linear model.

Suppose that some new variable, z_i, is suggested as being important, this is checked by plotting $\hat{e}_i$ against z_i. If this shows a linear relation we would have as our residual model

$$\hat{e}_i = \beta_2 z_i + f_i$$

where z_i is measured from its mean value (so $\bar{z} = 0$) to avoid interfering with our previous estimate of the mean of y. If our estimated slope of β_2 is say 6

$$\hat{e}_i = 6z_i + f_i$$

and so

$$y_i = 22 + 25x_i + 6z_i + f_i.$$

It must be emphasized that the procedure used in residual modelling does not give rise to optimum fitted models in any sense. It is a simple procedure for obtaining a first iteration of the original model. Even where two models fitted by least squares are used, the combined fitted model is not the same as that which would have been obtained by Least Squares applied to the combined model. The reason is simply that the original parameters, 22 and 25 in our example, were estimated by Least Squares, with the usual assumptions which we now know to be wrong. We therefore need to improve on their first iteration. This is done by using the form of the iterated model but not its estimated parameters. Thus in the previous examples we would consider

$$y_i = \beta_0 + \beta_1 x_i + \beta_2 x_i^2 + f_i$$
$$y_i = \beta_0 + \beta_1 x_i + f_i - \theta f_{i-1}$$
$$y_i = \phi y_{i-1} + \beta_0 + \beta_1 x_i + \beta_2 x_{i-1}1 + f_i$$
$$y_i' = \beta_1 + \beta_0 x_i' + f_i \qquad \left(y_i' = \frac{y_i}{x_i}, \ x_i' = \frac{1}{x_i} \right)$$
$$y_i = \beta_0 + \beta_1 x_i + \beta_2 z_i + f_i$$

and then apply Least Squares or some other method to estimate the parameters. The random variables f_i are now assumed to satisfy the standard assumptions, so standard methods can be applied. Notice that having done this we can evaluate the new residuals $\hat{f}_i$ and begin again to investigate whether they have any remaining structure.

As a final illustration we return to a problem discussed in subsection 16.5.2. The long-term steady state density for a species of larvae was modelled by

$$\ln N = \alpha(1 - \beta).$$

An investigation with a few numerical values will show that N is highly sensitive to changes in α and β. This corresponds to the possibility of stable situations where the larvae are rare and others where there is an infestation that can destroy the habitat. To study what factors might be relevant the model was fitted over a number of years and the residuals plotted against possible factors. It was found that quite small deviations from the normal temperature, $T - T_n$, at a certain time of the year showed a clear linear relation with the residuals thus

$$\hat{e} \simeq \alpha_1 (T - T_n) + f.$$

Hence in effect the parameter α became a linear function of $T - T_n$, and the model

$$\ln N_t = \alpha + \beta \ln N_{t-1} + e_t,$$

became

$$\ln N_t = \alpha_0 + \alpha_1 (T - T_n) + \beta \ln N_{t-1} + f_t.$$

20.3 Parameter modelling

It sometimes happens that a closer examination of the parameters leads to an elaboration of a model. For example, a parameter in a regression model represents how a regressor variable influences the dependent variable. Its constancy may depend on the constancy of the environment of the prototype. In practice the relationship of cause and effect may be changing in time or it may involve other variables whose importance is found during the validation. A natural response to such situations is to model the original parameter. This section illustrates this approach with some examples.

Example 1

A model for the seasonal variation, S, of an economic variable makes use of the addition of a number of sine and cosine curves. For example, we might have

$$S_t = \alpha_0 + \alpha_1 \sin \lambda t + \beta_1 \cos \lambda t + \alpha_2 \sin 2\lambda t + \beta_2 \cos 2\lambda t.$$

with $\lambda = \dfrac{2\pi}{12}$ to give a 12-month cycle.

Standard methods provide estimates of the parameters. Evidence from residual analysis often suggests that the seasonal pattern changes slowly over time. A parameter modelling approach may be adopted by using a moving section of data, i.e. first $S_1, \ldots, S_k$ then $S_2, \ldots, S_{k+1}$, $S_3, \ldots, S_{k+2}$, etc. For each section the parameters are estimated giving $\hat{\alpha}_{0i}, \hat{\alpha}_{1i}, \hat{\beta}_{1i}, \hat{\alpha}_{2i}, \hat{\beta}_{2i}$ for the ith section. The estimators are then plotted against i to see if there is any clear structure. If, for example $\hat{\alpha}_{1i}$ shows a trend then we might model it by

$$\hat{\alpha}_{1i} = \hat{\alpha}_{11} + \hat{\gamma}_1 i,$$

a fit by eye providing initial estimators. The revised model thus has terms

$$\hat{\alpha}_{12} \sin \lambda t + \hat{\gamma}_1 t \sin \lambda t.$$

Note that $\hat{\alpha}_{11}$ has to be adjusted to enable the origin of the i variables to align with that of the t variables to give the $t \sin \lambda t$ term. The reference by Nettheim details this situation.

In the application of Bayes formula we have often used the prior distributions of parameters. Often these priors have been chosen on the basis of theoretical considerations. This is effectively conceptual parameter modelling. In some situations a much more empirical approach is possible.

Example 2
Records show that bottles of chemical obtained from a single production run show a percentage impurity that is $N(\mu, \sigma^2)$. However, different production runs give rise to different μ. Over a number of runs it is established that μ is itself Normally distributed. We thus have an empirically obtained prior distribution of the parameter μ.

Example 3
We have referred several times to the plotting of residuals against new variables. Suppose that in some situation the appropriate model is

$$y_i = \alpha + \beta z_i x_i + e_i$$

whereas the model

$$y_i = \alpha + \beta^* x_i + e_i^*$$

is used. The residuals from the fitted model will be approximately

$$\hat{e}_i^* \simeq (\beta z_i - \beta^*) x_i + e_i.$$

Thus a plot of $\hat{e}_i^*$ against z_i may not clearly reveal the influence of z, owing to the multiplying effect of x. Suppose, however, we subdivide the data into groups that have approximately the same z, and for each group obtain the estimator $\hat{\beta}_z^*$. A plot of $\hat{\beta}_z^*$ against z will correspond to the model

$$\beta_z^* = \beta z$$

and should show up the relationship clearly.

20.4 Variable modelling

A careful study of the variables used in a well validated model may sometimes suggest alternatives that simplify or clarify the model. We have termed this process variable modelling. Again let us consider some examples.

Example 1
An empirical study of the 'strength' of wooden logs led to a best fitting model of

$$\ln s = 1.3 + 1.1 \ln l + 1.95 \ln d$$

where $l =$ length of log, $d =$ diameter of log.
This gives $s = $ const. $l^{1.1} d^{1.95}$.

Approximating the log's shape by a cylinder we get a volume of const. $l d^2 = v$. This was within the margin of error of the coefficients in the orginal model and thus

$$s = \beta V = \beta l d^2$$

is suggested as a simpler and more natural form of model.

Example 2

Beltrami considers a model for the time police cars take to get to the sites of incidents. Let the site of the incident be regarded as the origin. Suppose a police car is at position (x, y) then we might use as its distance the euclidean straight line distance obtained by Pythagoras

$$d = \sqrt{x^2 + y^2}.$$

However, if we examine the model of the situation more carefully we might be nearer the truth, certainly in an American context, if we consider a rectangular array of streets as in Figure 20.1. The road distance the car has to cover is

$$d_1 = |x| + |y|.$$

This way of measuring is referred to as the *Manhattan Metric*. Two features can be seen from the diagram. First that the distance d_1 is less than l for all points in the triangle OAB. Thus if cars are distributed at random with density γ cars per unit area then the expected number of cars within l units of O (using the d_1 metric and all four quadrants round O) is $4.\gamma.\frac{1}{2}l^2 = 2\gamma l^2$.

From the Poisson model we could then argue that the probability of there being no car within l is $\exp -2\gamma l^2$. The second feature of note is that comparing d_1 with d the maximum difference will occur along the diagonal OP and they will be equal along the axes OA and OB. To consider the relation further write

$$d_1 = \theta d$$

where θ obviously depends on x and y, in fact

$$\theta^2 = 1 + \frac{2xy}{x^2 + y^2}.$$

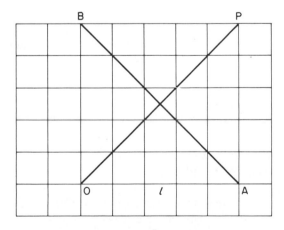

Figure 20.1 The Manhattan metric

This gives $\theta = 1$ along the axes and $\theta = \sqrt{2}$ along the diagonal so $1 \leqslant \theta \leqslant \sqrt{2}$ for a Manhattan street layout. If we assume a random position for a car with X and Y independently distributed on $(0, 1)$ for convenience it is found that the expectation of θ is approximately 1.2 which gives an interesting contrast with the empirical figure of 1.3 for the Sheffield data of Chapter 1.

Example 3

A common use of variable modelling occurs where sophisticated models devised for situations in the developed world, where adequate data is available, are to be applied in situations in the developing world where data is not so readily available. An example is the modelling of drainage basins, where the necessary detailed rainfall data is lacking in many countries. Thus rainfall, x_{ij}, in a sub region i for week j will be required for the model but will not be recorded. The procedure adopted is to model x_{ij} using other relevant and available information. The x_{ij} for several sub regions with known rainfall might be regressed on some physical characteristics of the regions, seasonal variables and national rainfall figures. The fitted regression models will then be used to generate values, $\hat{x}_{ij}$, that will be used in the model. Examples of such procedures are given in Carey and Haan and Jarboe and Haan.

20.5 Substructure modelling*

In Chapter 18 we mentioned the use of models in choosing between theories, and pointed out that several theories could lead to the same model. In the present context we may similarly take an empirically derived model and enquire what substructures, theories, could have led to such a structure. There may well be alternative substructures that lead to the same overall behaviour. To choose between these will require observation of the variables in the substructure. None the less the attempt to build explanatory substructures helps considerably in the understanding of model behaviour.

Example 1

Consider the simple Autoregressive, AR(1), model

$$x_t = \theta x_{t-1} + e_t.$$

This is a very common model and one can envisage a number of possible substructures that could lead to its occurrence:

(a) Another variable z may interrelate linearly with x_t in such a way that causal effects take $\frac{1}{2}$ a time unit to take place. Thus

$$x_t = \alpha z_{t-\frac{1}{2}} + u_t, \qquad z_t = \beta x_{t-\frac{1}{2}} + v_t$$

where u and v are random terms.

Combining gives

$$x_t = \alpha\beta x_{t-1} + (u_t + \alpha v_{t-\frac{1}{2}}),$$

which has AR(1) form.

(b) Suppose that x_t depends on all past z on the form of a geometrically distributed lag model (see Chapter 7.2.2) with a unit cause and effect delay. Thus

$$x_t = z_{t-1} + \alpha z_{t-2} + \alpha^2 z_{t-3} + \ldots + e_t$$

It follows that

$$x_t = z_{t-1} + \alpha(z_{t-2} + \alpha z_{t-3} + \ldots) + e_t$$
$$= z_{t-1} + \alpha x_{t-1} + e'_t$$

If α is close to one and the x and z are measured from their mean levels, the term z_{t-1} makes a relatively small and possibly random contribution which can be absorbed into the error term giving, approximately, the AR(1) model

$$x_t = \alpha x_{t-1} + e_t.$$

(c) Consider a situation where it is possible to influence the variable x and where there is some target value that is optimum. Let the scale be chosen to make the target value zero. A control adjustment, z, will be made to the situation which will depend on how far the most recent observation is from the target. Thus

$$z = -k x_{t-1}.$$

Suppose the adjustment alters the current value, reducing it for positive x. The future value will also be subject to error, so

$$x_t = x_{t-1} + \beta z + e_t$$

Hence

$$x_t = (1 - \beta k)x_{t-1} + e_t$$

which again is AR(1).

The reference by Young gives a discussion of this topic for more general models.

Chapter 21

Improving estimation*

21.1 Reducing bias

In the first iteration of a modelling exercise the methods used are those that are simplest in terms of the model or are at least those of standard application such as maximum likelihood. On further iterations we may wish to obtain improved estimators with properties that are more appropriate for the particular application proposed. We may for example seek to reduce the bias in biased estimators. If interest focusses on a specific parameter value we may try to reduce the Mean Square Error of its estimator. Considering bias both on its own and as a contribution to the Mean Square Error suggests that adjustments that can reduce the bias would be well worth considering.

21.1.1 Jackknife estimation

One procedure for reducing bias, referred to as 'Jackknife estimation', is based on the approach of *Sample Splitting*. In sample splitting we divide our sample of data into two parts and do some appropriate calculations on one part. This process is repeated using different divisions of the data. Finally all the results are combined. In 'Jackknife estimation' the 'split' consists of simply removing one observation, say x_j. From the remaining $n - 1$ observations we calculate the estimator of parameter, θ, which we will denote by $\hat{\theta}_{-j}$. This is repeated for all possible splits giving $\hat{\theta}_{-1}$, $\hat{\theta}_{-2}, \ldots, \hat{\theta}_{-n}$. It is usually the case that when the bias of an estimator is suitably expressed for large n the dominant term takes the form c/n, where c may depend on θ but not on the number of observations n. Thus the bias of $\hat{\theta}$ will be approximately c/n and of $\hat{\theta}_{-j}$ will be approximately $c/(n-1)$. It follows that the estimator $\hat{\theta}_*$, where

$$\hat{\theta}_* = n\hat{\theta} - \frac{(n-1)}{n} \sum_j \hat{\theta}_{-j},$$

302

will have a bias, for large n, of

$$b(\hat{\theta}_*) \simeq c - \frac{1}{n}\sum_1^n c = 0.$$

Thus $\hat{\theta}_*$ will be approximately unbiased. For illustration Table 2.1 shows the calculation of a Jackknifed estimator of λ for a small sample from an exponential distribution. The value of $\hat{\lambda}$ obtained underlines the potential hazard of using large sample methods with small samples.

21.1.2 Approximation methods

An alternative approach that can sometimes be used is based on Taylor series (Appendix 2). We have already used the first two terms to linearize models in Chapter 13. We now extend the Taylor series to the next term giving

$$g(z) \simeq g(\theta) + (z-\theta)\left(\frac{dg}{dz}\right)_\theta + \tfrac{1}{2}(z-\theta)^2\left(\frac{d^2g}{dz^2}\right)_\theta.$$

Now choose variable z and parameter θ so that the term $z-\theta$ has zero expectation. In this exponential example let $z = \bar{x}$ and $\theta = 1/\lambda$. As our first estimator of λ is $1/\bar{x}$ we let $g(z) = 1/z$ and hence

$$\frac{dg}{dz} = -\frac{1}{z^2} \text{ and } \frac{d^2g}{dz^2} = \frac{2}{z^3}.$$

The Taylor series thus becomes

$$\frac{1}{\bar{x}} \simeq \lambda + \left(\bar{x} - \frac{1}{\lambda}\right)(-\lambda^2) + \tfrac{1}{2}\left(\bar{x} - \frac{1}{\lambda}\right)^2(2\lambda^3).$$

Taking expectations of both sides and noting that $E\left(\bar{x} - \frac{1}{\lambda}\right) = 0$ and $E\left[\left(\bar{x} - \frac{1}{\lambda}\right)^2\right] = V(\bar{X}) = \frac{1}{n\lambda^2}$ gives

$$E\left(\frac{1}{\bar{x}}\right) = \lambda + 0 + \tfrac{1}{2} \cdot \frac{1}{n\lambda^2} \cdot 2\lambda^3 = \lambda + \lambda/n.$$

The adjustment term $\tfrac{1}{2}(z-\theta)^2 g''(\theta)$ is thus λ/n. The general procedure is to estimate this term and use it to adjust the left-hand side. λ/n is estimated by $1/(n\bar{x})$ and thus the adjusted estimator is

$$\frac{1}{\bar{x}} - \frac{1}{n\bar{x}} = \frac{n-1}{\sum x}.$$

This will have expectation of approximately λ. In actual fact if the distribution of $1/\sum x$ is derived it is found that this estimator is exactly unbiased. Table 21.1 illustrates this formula.

Table 21.1 Jackknife and adjusted estimators for exponential data (10 observations)

x_j	$\Sigma x - x_j$	$\hat{\lambda}_{-j} = \dfrac{9}{\Sigma x - x_j}$	
36.1	49.1	0.183	(i) Maximum Likelihood Estimator
8.8	76.4	0.118	$\hat{\lambda}_1 = \dfrac{10}{85.2} = 0.117$
5.3	79.9	0.113	
1.5	83.7	0.107	(ii) Jackknife Estimator
0.1	85.1	0.106	$\hat{\lambda}_2 = 10\,(0.117) - \dfrac{9}{10}\,(1.204)$
1.6	83.6	0.108	
2.3	82.9	0.109	
17.6	67.6	0.133	$= 0.087$
7.1	78.1	0.115	(iii) Taylor Series adjustment
4.8	80.4	0.112	$\hat{\lambda}_3 = \dfrac{9}{85.2} = 0.106$
$\Sigma x = 85.2$		1.204	

21.1.3 Two-stage least squares

We saw in subsection 13.3.2 that a correlation between regressor and error variable can lead to biased estimators of parameters. The method of Two-Stage Least Squares is designed to produce asymptotically unbiased and consistent estimators in spite of such correlations. Its main area of application is in simultaneous equation models.

Example 2
Consider again the simultaneous equation model of subsection 7.2.4 and in particular the equation

$$i_t = \beta_0 + \beta_1 y_t + \beta_2 y_{t-1} + e_{2t}$$

Examining Figure 7.6(c) shows that y_t is influenced by e_{2t} and so we expect a correlation and doubt that the assumptions for Least Squares are valid. The approach to the solution is equivalent to the instrumental variable method in that we will replace y_t by a proxy variable with the same expectation, so that β_1 is unaltered, but with less correlation with e_{2t}. The method is to use the Reduced Form equation for y_t to get a fitted value $\hat{y}_t$. The Reduced Form is

$$y_t = \pi_{30} + \pi_{31} g_t + \pi_{32} y_{t-1} + f_{3t}.$$

We can fit this directly by least squares to get

$$\hat{y}_t = \hat{\pi}_{30} + \hat{\pi}_{31} g_t + \hat{\pi}_{32} y_{t-1}$$

and

$$y_t = \hat{y}_t + \hat{f}_{3t}.$$

The residual $\hat{f}_{3t}$ is an estimate of f_{3t} and so the influence of e_{2t} on y_t will be picked up largely by $\hat{f}_{3t}$ leaving $\hat{y}_t$ relatively uncorrelated with e_{2t}. It will not be exactly uncorrelated, since it depends on the estimates $\hat{\pi}$ which depend on y_t and hence on e_{2t}. Nonetheless the effect will be small and tend to zero as more data leads to better estimates of the π. The fitted $\hat{y}_t$ is now used as a proxy for y_t to create the new model

$$i_t = \beta_0 + \beta_1 \hat{y}_t + \beta_2 y_{t-1} + e_{2t}^*.$$

This model can now be fitted by Least Squares to give estimates that will be satisfactory for large sets of data.

21.2 Reducing mean square error

Suppose we have an unbiased estimator $\hat{\theta}$ of θ whose variance is σ^2. The mean square error (MSE $= E(\theta - \hat{\theta})^2$) is equal to σ^2. The question arises whether we can adjust $\hat{\theta}$ to get an estimator with smaller MSE. The simplest adjustment is to multiply $\hat{\theta}$ by some quantity, a, to get an estimator $\hat{\theta}_1 = a\hat{\theta}$.

Now MSE = Variance + bias^2 (see section 14.4), and

$$V(\hat{\theta}_1) = a^2 \sigma^2,$$
$$b(\hat{\theta}_1) = E(a\hat{\theta}) - \theta = (a - 1)\theta.$$

Hence MSE $(\hat{\theta}_1) = a^2 \sigma^2 + (a - 1)^2 \theta^2$.

This is minimum when a is

$$a^* = \frac{\theta^2}{\sigma^2 + \theta^2}.$$

Noting that the coefficient of variation of $\hat{\theta}$ is $v = \sigma/\theta$ we have

$$\hat{\theta}_1^* = a^* \hat{\theta} = \hat{\theta}/(1 + v^2).$$

Substituting will show that

$$\text{MSE } (\hat{\theta}_1^*) = a^* \sigma^2.$$

As a^* is less than one it can be seen that by this simple adjustment we have improved the MSE property of the estimator. Unfortunately the estimator $a^* \hat{\theta}$ is a theoretical entity, as it depends on knowing the coefficient of variation v, which formally depends on both θ and σ^2. However, there are some practical situations where there are ways of finding v without knowing either θ or σ^2.

Example 1
Consider a Normal Distribution with data $x_1, \ldots, x_n$. We usually estimate σ^2 by

$$s^2 = \frac{1}{n-1} \sum (x_1 - \bar{x})^2.$$

This is an unbiased estimator of σ^2 which can be shown to have variance $2\sigma^4/(n-1)$. Hence $v^2 = 2/(n-1)$ which is independent of σ. Substituting gives $a^* = (n-1)/(n+1)$ so the adjusted estimator is

$$s^{*2} = \frac{1}{n+1} \sum (x_i - \bar{x})^2.$$

As a^* indicates the magnitude of the reduction in MSE, it is evident that for small samples an appreciable reduction is obtained, but that for large n there is little difference between s^2 and s^{*2}.

21.3 Identifying and reducing loss

In the discussion both of estimation and of decision making we have referred to the criteria used. We have considered residual-based criteria, such as $\sum \hat{e}^2$, mean square error criteria such as $E[(\hat{\theta} - \theta)^2]$ and general utility or loss functions giving the cost of decisions made in relation to the 'state of nature'. We have throughout assumed that we knew what these criteria were. They were conceptually given functions. In reality we have to decide what criteria to use. Suppose for example an error, e, refers to the difference between the number of manufactured units ordered and the number produced. In this case it is by no means obvious what function of e would be appropriate. A non-symmetrical function is suggested. A simple model for such a function is the *Linex Loss Function* defined by

$$L(e_i) = b \exp(ae_i) - ce_i - b \qquad a, b, c > 0$$

This is illustrated in Table 21.2. We are immediately faced with questions of whether this is the appropriate model, a question of identification, and if so what values to assign the parameters (a, b, c), a question of estimation.

Table 21.2 Illustrative values of the Linex Loss Function
$L(e) = 2\exp(2e) - 10e - 2$

e	$L(e)$	
-5	48.00	
-4	38.00	↑ linear increase
-3	28.01	
-2	18.04	
-1	8.27	
0	0	
1	2.78	
2	87.20	
3	774.86	↓ exponential increase
4	5919.21	
5	44000.96	

It is evident from this example that practical problems sometimes raise the need to identify and fit loss functions themselves as part of the modelling process. Obviously to do this a model of the prototype has to be fitted by some method to generate a set of values for $\hat{e}_i$. The consequent loss $L(\hat{e}_i)$ has then to be measured for the problem in hand and the loss model identified and fitted. Having done this we can then return to the original model for the prototype, and fit again using the new criteria. The processes of identifying and fitting models for the prototype and the loss form a process of iteration, each in turn being fixed whilst the other is revised.

In the previous section we used an estimator criterion $E[(\theta - \hat{\theta})^2]$, where the expectation was over the distribution of the data random variable. If there is prior information available about the parameter in the form of a prior distribution, even if it is a vague prior, then we can sometimes make progress with very difficult problems.

Example 1

Consider a problem that at first sight is quite straightforward. The observations x are the time taken to cover unit distance by a vehicle. It is assumed as a reasonable model that $X \cap N(1/v, \sigma^2)$ where $1/v$ is the mean time, v the velocity and the error is assumed to arise solely in the measurement of time. A natural estimator of v is $1/\bar{x}$. However, the fact that the normal distribution includes the zero point has the effect that $1/\bar{x}$ has none of the usual properties of good estimators. We thus have the problem first met in the example of calibration. Suppose now we consider an estimator $\hat{v}$ of v and measure the quality of this estimator by the loss function

$$l(v, \hat{v}) = E\left[\left(\frac{v - \hat{v}}{v} \right)^2 \right],$$

where the expectation is calculated for the posterior distribution of v. The estimator $\hat{v}$ must be a positive value so the loss function stays finite over the distribution of $v > 0$. This corresponds to sensible values given our prior knowledge of the situation.

To go from prior to posterior distributions it is convenient to have a vague prior for the mean time $\theta = 1/v$. If $\hat{v}$ is not close to the origin there is little difference in effect between the use of a vague prior for v or for θ. The loss function can be re-written as

$$l(v, \hat{v}) = E[(1 - \theta\hat{v})^2],$$

where the expectation is with respect to the posterior distribution of θ.

Differentiating with respect to $\hat{v}$ to minimize gives

$$-2E(\theta) + 2\hat{v}E(\theta^2) = 0,$$

hence $\hat{v} = E(\theta)/E(\theta^2) = E(\theta)/[E(\theta)^2 + V(\theta)]$.

A suitable prior is the normal prior of Table 15.1, 3, with $m = 0$. For this prior and normal data the posterior distribution of θ given the data is $N(\bar{x}, \sigma^2/n)$. Hence

$$\hat{v} = \frac{\bar{x}}{\bar{x}^2 + \dfrac{\sigma^2}{n}}.$$

This can be re-written as

$$\hat{v} = \frac{1}{\bar{x}}\left(\frac{1}{1 + 1/u^2}\right),$$

where $u = \dfrac{\bar{x} - 0}{\sigma/\sqrt{n}}$, which is the statistic that is used to test the deviation of the mean from zero. When u is small the bracketed term is also small and the estimator is reduced from $1/\bar{x}$. Where $\bar{x}$ is a long way from zero the bracketed term tends to one so that $1/\bar{x}$ is the value used. We thus obtain a *Shrinkage Effect* similar to that produced in the previous section. In a number of areas an improved estimator can be obtained by shrinking a standard estimator towards zero or towards some prior mean.

21.4 Reducing the effects of outliers

An outlier is an observation which is sufficiently extreme that it can be taken as being caused by some undesired factor not allowed for in the model, for example, an error in recording or in observation. Such errors may leave an observation looking much like its fellows and it will be unrecognizable unless other evidence points to its unreliability. We will thus usually only detect outliers that show distinctly different behaviour from the other observations. It has happened that such behaviour was so clear and the factor producing the outliers so evident that, rather than leaving out the observation, the factor was simply added to the model. In general, however, the occurrence of outliers leaves only two options. The first is to seek to identify the outliers, remove them from the data and repeat the modelling process. The second is to use methods designed to minimize the effects of any outliers. Which option is used depends on the amount of data available and the aims of the modelling exercise. If only a small amount of data is available then a study of the data and a rejection of outliers is feasible. If large amounts of data are being used it becomes almost certain that the data will contain outliers and their likely effects should be considered. In such circumstances methods that are insensitive to the presence of outliers, usually called *Robust* methods, should be considered, see the reference by Launer and Wilkinson.

Example 1
In the data on road and straight line distances for Sheffield, a first plot showed one observation that was a long way from the straight line. The observation was dropped from the data and the line fitted. The standard

error of the residuals was calculated and the error for the rejected observation was found to be about four standard deviations. Thus statistically the decision to drop the outlier was supported. Outliers should not, however, be dropped without some attempt to ask the question why they occurred. In this case the reason was found to be the occurrence in Sheffield of a number of long cliffs that form natural barriers. Thus one may have to drive two miles to get to a point a half-a-mile away. The implication of dropping the outlier is that as a first attempt at modelling the prototype only includes journeys that do not involve detours round these cliffs. If such journeys are to be included a more elaborate model would be required.

Example 2
In fitting the line to the data of the car distance example the least squares estimator was used:

$$\hat{\theta} = \sum_{i=1}^{n} x_i y_i / \sum_{i=1}^{n} x_i^2.$$

Consider the effect of just one of the y values, y_1, on the estimator $\hat{\theta}$.

$$\hat{\theta} = \sum_{i=2}^{n} x_i y_i / \sum_{i=1}^{n} x_i^2 + y_1 (x_1 / \sum_{i=1}^{n} x_i^2)$$

writing $y_1 = \theta x + e_1$ and finding the expectation of $\hat{\theta}$, treating only $y_2 \ldots y_n$ as random variables, gives

$$E(\hat{\theta}|e_1) = \theta + e_1(x_1 / \sum x_i^2)$$

If y_1 is an outlier e_1 will be large and its influence on $\hat{\theta}$ will depend on a factor $x_1 / \sum x_i^2$. There is thus a linear influence of e_1 on $\hat{\theta}$, as shown in Figure 21.1. The plot of Figure 21.1(a) is called an *Influence Curve*. Notice that in this regression situation the influence of an error depends on both its magnitude and its regressor variable. If we are to handle outliers effectively we would want to stop the influence of e_1 increasing linearly. We might for example put some maximum limit on it, so errors greater than e_{max} are ignored, as in Figure 21.1(b). Alternatively we might make the curve smoothly tend to zero for very large errors, which we would effectively reject, as in Figure 21.1(c). In practice estimators with these properties are obtained by introducing a weighting, w_i, in the Least Squares criterion and consequently in the summations derived from it. So we would minimize $\sum w_i \hat{e}_i^2$ to obtain

$$\hat{\theta} = \sum w_i x_i y_i / \sum w_i x_i^2.$$

The influence curve of Figure 21.1(b) is achieved by setting $w_i = 1$ for $\hat{e}_i \leqslant c$ and zero otherwise, where $\hat{e}_i$ is the corresponding residual from a previous fit and the constant c is fixed as a suitable multiple of the estimated standard deviation, say between 2 and 3. Having re-estimated $\hat{\theta}$ a new set of

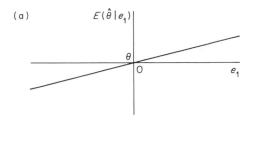

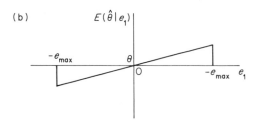

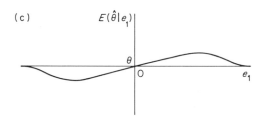

Figure 21.1 Influence curves

residuals can be found and hence a new set of weights. The sequence of operations is repeated until a stable $\hat{\theta}$ is obtained. The curve of Figure 21.1(c) is obtained by using the *Biweight*:

$$w_i = \begin{array}{ll} (1-u_i^2)^2 & |u| \leqslant 1 \\ 0 & \text{otherwise} \end{array}$$

where $u_i = \hat{e}_i/2c$ and c is as defined above. The divisor $2c$ ensures that the condition $|u_i| \leqslant 1$ will normally hold. Again an iterative approach is used. It should be noted that this approach not only deals with outliers, but also cuts down the general over-emphasis of observations with large residuals, that we noted as a possible weakness of the method of least squares.

21.5 Local fitting

The approximate and limited nature of models has been frequently emphasized in foregoing pages. Often we shall be able to specify the limitation explicitly, for example that the model only applies to a particular locality in time or space. If this is the case then it is clearly not sensible to use data from outside this locality in estimating the model parameters. With progression models the locality is not always clearly defined but usually the present represents the main time of interest.

Example 1
We return to the model of a local linear trend (Example 2 of section 19.3)

$$x_t = \mu_t + e_t,$$
$$\mu_t = \mu_{t-1} + \beta_{t-1} + u_t,$$
$$\beta_t = \beta_{t-1} + v_t.$$

At any time t interest focusses on estimating the current mean and slope μ_t, β_t. A way of obtaining such 'local estimates' is provided by the idea of smoothing, introduced in the discussion on monitoring in section 17.6. The smoothed average of observations z_t is obtained from

$$\tilde{z}_t = S_t(z)/W_t$$

Where

$$S_t(z) = z_t + a z_{t-1} + a^2 z_{t-2} + \ldots, \quad 0 < a < 1,$$

and

$$W_t = 1 + a + a^2 + \ldots.$$

If there is sufficient data the series for W sums to $1/(1-a)$. The weighted sum can be written as

$$S_t(z) = z_t + a S_{t-1}(z)$$

and so

$$\tilde{z}_t = (1-a)z_t + a\tilde{z}_{t-1}.$$

Thus if a is close to unity each new estimate $\tilde{z}_t$ puts a small proportion of the new observation z_t with the appropriate large proportion of the previous $\tilde{z}_{t-1}$. In words we have:
new estimate $= (1-a)$(new data) $+ a$(old estimate).
   Applying this to the estimate of the current mean, $\hat{\mu}_t$ we have

$$\hat{\mu}_t = (1-a)x_t + a(\hat{\mu}_{t-1} + \hat{\beta}_{t-1}),$$

since the estimate of μ_t based on information available at time $t-1$ was $\hat{\mu}_{t-1} + \hat{\beta}_{t-1}$. The latest evidence available on the slope is represented by $\hat{\mu}_t - \hat{\mu}_{t-1}$ and so the updating formula for the estimate of β_t is

$$\hat{\beta}_t = (1-a^*)(\hat{\mu}_t - \hat{\mu}_{t-1}) + a^* \hat{\beta}_{t-1}$$

where a^* is a different constant used for smoothing the slope. It will be seen that the estimates of μ and β take the form of recurrence relations that are updated as each new observation is obtained to provide estimates of the current μ_t and β_t. The *Smoothing Parameters*, a and a^*, control how much emphasis is placed on the new and on the old information.

With methods such as the one illustrated the process of estimation is seen as inherently an iterative process. The model is not fitted as a once-and-for-all act, but is continually refitted as new data becomes available. If classical methods of estimation are used in such situations more and more data becomes available and less and less notice is taken of new data as it arrives. Intuition suggests that in an unstable world this is not a wise approach and it would be better to allow for drifting of parameter values.

Classicial methods of estimation can be modified by introducing weights, such as the exponential weights, a^r, used here. These methods then lead to estimators involving weighted averages such that recent data is emphasized and distant past data forgotten. These methods are discussed by Gilchrist (References to Part VI). The method described in the example is a generalization of the form taken, with a long series of observations, by the method of least squares when exponential weights are introduced.

21.6 Postscript

Let us finish this chapter, and this book, by taking a brief look back at the main questions that face the builder of statistically based models.

(a) *Why is the model required?*

This is not a statistical question but a knowledge of the answer is essential in providing a framework in which to consider the more statistical aspects of modelling. As we saw in Chapter 18 the intended application influences the decisions made at all the stages of modelling. In Section 3 of this Chapter we saw that when we move in the iteration process to new levels of sophistication it is the application that guides the choice of criteria.

(b) *What information and resources are available or obtainable?*

We have repeatedly emphasized the differences between conceptual and empirical approaches at all stages in the modelling process. We have also indicated the weaknesses of each separately and the value in most situations of bringing the two together in a more eclectic approach. To do this we need to create an information base of both knowledge and data. This provides the necessary resource for identification, estimation and validation. We must not forget that, in relation to the question of (a), the model-building phase will be followed by a model-using stage. This too will require its own information and resources. What information and resources will be available at both stages needs consideration at the very outset of the modelling exercise. It is pointless ending the exercise with a sophisticated model on an organization's computer if the expertise required to use it effectively is not available in the organization, or if the quality of data needed to get sensible behaviour from the model is not obtainable.

(c) What models should be used?

Choosing a model is rather like choosing a computer. It is very attractive, if one can afford it, to go for the most sophisticated model available. It looks good, it provides lots of options, it impresses your friends. However, experience in both modelling and computing indicate that such situations can often lead to mistakes and difficulties. These failures in turn lead to neglect of even the good and simple features available. It is far better to apply the principle of simplicity and iterate slowly as experience and confidence builds up. There is evidence that there is often a law of diminishing returns in modelling. The major benefits come from the thought and effort required to build and use the simplest models. The benefits of moving to better models of the prototype may be marginal and certainly need to be weighed up before taking this step.

(d) What approach should be taken?

There has been an emphasis in statistics, quite naturally, on an empirical, data based, approach to modelling. Unfortunately this has often led to a totally unbalanced approach in which the meaning and background of the data has been totally ignored. Statistical modelling takes place at the interface between the methodologies provided by statistics and the specific problem area from science, social science, business, etc. for which the model is required. As will have been evident throughout this book, there are no universal techniques for working at this interface, no simple ways of merging the empirical and conceptual approaches. There has been a great deal of ad hockery about our discussions of eclectic approaches. Indeed, the eclectic approach often involves separately taking both empirical and conceptual approaches and hoping that the results can be intuitively combined. None the less it is the thesis of this book that successful modelling requires the bringing together of the empirical and the conceptual, or, in human terms, the co-operation of the statistician and the 'owner' of the problem that requires the model.

References and reading list

Part I Introduction

Aris, R. (1978). *Mathematical Modelling Techniques*. Pitman. London.

Freudenthal, H. (Ed.). (1961). *The Concept and the Role of the Model in Mathematics and the Social Sciences*. Reidel. Dordrecht.

Jacoby, S. L. S., and Kowalik, J. S. (1980). *Mathematical Modelling with Computers*. Prentice-Hall. Englewood Cliffs, New Jersey.

Klir, J. and Valach, M. (1966). *Cybernetic Modelling*. Iliffe Books. London.

Koutsoyiannis, A. (1977). *Theory of Econometrics*. 2nd edn. Macmillan Press. London.

Lancaster, P. (1976). *Mathematical Models of the Real World*. Prentice-Hall. Englewood Cliffs, New Jersey.

Orlob, G. T. (1975). Present problems and future prospects of ecological modelling. *Ecological Modelling*. Resources for the Future Inc. Washington D.C.

Wold, H. O. (1968). Model building and scientific method. A graphic introduction, in *Mathematics Model Building in Economics and Industry*. First series. CECR. Charles Griffin. London.

Zeigler, B. P. (1976). *Theory of Modelling and Simulation*. John Wiley. New York.

Zeigler, B. P. (Ed.). (1979). *Methodology in Systems Modelling and Simulation*. North-Holland. Amsterdam.

Part II Common statistical models

Bondell, M., Gilmour, D. A., and Sinclair, D. F. (1979). A statistical method for modelling the flow of rainfall in a tropical rain forest catchment. Journ. of Hydrology, **42**, 251–267.

Box, G. E. P., and Jenkins, G. M. (1976). *Time Series Analysis: Forecasting and Control*. Holden-Day. San Francisco.

Doebelin, E. O. (1980). *System Modelling and Response*. John Wiley. New York.

Draper, N. R., and Smith, H. (1981). *Applied Regression Analysis*. 2nd edn. John Wiley. New York.

Duncan, O. D. (1975). *Introduction to Structural Equation Models*. Academic Press. London.

Forrester, J. W. (1968). *Principles of Systems*. MIT Press. Cambridge, Mass.

Freund, J. E., and Walpole, R. E. (1980). *Mathematical Statistics*. 3rd edn. Prentice-Hall. London.

Hahn, G. J., and Shapiro, S. S. (1967). *Statistical Models in Engineering*. John Wiley. New York.

Harvey, A. C. (1981). *Time Series Models*. Philip Allen. Oxford.

Harvey, A. C. (1981). *The Econometric Analysis of Time Series*. Philip Allen. Oxford.

Hastings, N. A. J., and Peacock, J. B. (1974). *Statistical Distributions*. John Wiley. New York.

Hoel, P. G. (1971). *Introduction to Mathematical Statistics*. 4th edn. John Wiley. New York.

Howard, R. A. (1971). *Dynamic Probabilistic Systems*. Vol I, *Markov Models*. John Wiley. New York.

Kashyap, R. L., and Rao, A. P. (1976). *Dynamic Stochastic Models from Empirical Data*. Academic Press. New York.

Kendall, M. G., and Stuart, A. (1976). *The Advanced Theory of Statistics*. 3rd edn. Charles Griffin. London.

Legusto, A. A., Forrester, J. W., and Lyneis, J. (1980). *Systems Dynamics*. American Elsevier. New York.

Meadows, D. L., and Meadows, D. H. (Ed.). (1973). *Toward Global Equilibrium*. Wright-Allen Press. Cambridge, Mass.

Neave, H. R. (1978). *Statistical Tables*. Allen & Unwin, London.

Nelder, J. A., and Wedderburn, R. W. M. (1972). Generalized linear models. *J. R. Statist. Soc. A*, **135**, Part 3, 370–384.

Nicholson, H. (Ed.). (1980/81). Modelling of Dynamic Systems. Vols I & II. I.E.E. Control Engineering Series. London.

O'Muircheartaigh, G. A., and Payne, C. (1977). *The Analysis of Survey Data*. Vol. 2. *Model Fitting*. John Wiley, Chichester, England.

Parzen, E. (1962). *Stochastic Processes. Holden-Day. San Francisco*.

Randers, J. (1980). *Elements of the System Dynamics Method*. MIT Press. Cambridge. Mass.

Renton, G. A. (Ed.) (1975). *Modelling the Economy*. Social Science Research Council. UK.

Seber, G. A. F. (1977). *Linear Regression Analysis*. John Wiley. New York.

Shapiro, S. and Gross, A. J. (1981). Statistical Modelling Techniques. Marcel Dekker. New York.

H. M. Treasury. (1979). Macroeconomic Model. Technical Manual. HMSO. London.

Upton, S. G. (1978). *The Analysis of Cross-tabulated Data*. John Wiley. New York.

Wetherill, G. B. (1972). Elementary Statistical Methods, 2nd edn. Chapman & Hall. London.

Part III Identification

Allen, D. M. (1971). Mean square error of prediction as a criterion for selecting variables. *Technometrics*, **13**, 469–475.

Atkinson, A. C. (1978). Posterior probabilities for choosing a regression model. *Biometrika*, **65**, 39–48.

Barry, C., and Wildt, A. (1977). Statistical model comparison in marketing research. *Management Science*, **24**, 387–392.

Bass, F. M. (1969). A new product model for growth of consumer durables. *Management Science*, **15**, 215–227.

Beachley, N. H., and Harrison, H. L. (1978). *Introduction to Dynamic Systems Analysis*. Harper & Row. New York.

Belsley, D. A., Kuh. E., and Welsch, R. E. (1980). *Regression Diagnostics*. John Wiley. New York.

Bergstrom, A. R. (1967). *The Construction and Use of Economic Models*. E.U.P. London.

Blattberg, R. C., and Sen, S. K. (1975). A Bayesian technique to discriminate between stochastic models of brand choice. *Management Science*, Series B, **21**, 682–696.

Box, G., and Hill, W. (1967). Discrimination among mechanistic models. *Technometrics*, **9**, 57–71.

Burghes, D. N., and Borrie, M. S. (1981). *Modelling with Differential Equations.* Ellis Horwood. Chichester, England.

Cox, D. R. (1962). Tests of Separate Families of Hypotheses. *J. Roy. Statist. Soc. B.* **24**, 406–424; and Proc. 4th Berk. Symp. Vol. 1, 105–123.

Davies, W. D. T. (1970). *System Identification for Self-Adaptive Control.* John Wiley. New York.

Doebelin, E. O. (1980). *System Modeling and Response.* John Wiley. New York.

Dym, C. L., and Ivey, E. S. (1980). *Principles of Mathematical Modelling.* Academic Press. New York.

Ehrenberg, A. S. C. (1975). *Data Reduction.* John Wiley. Chichester, England.

Ehrenberg, A. S. C. (1982). *A Primer on Data Reduction.* John Wiley. Chichester, England.

Eykhoff, P. (1974). *System Identification.* John Wiley. New York.

Gaver, K. M., and Geisel, M. S. (1974). Discriminating among alternative models: Bayesian and Non-Bayesian Methods. Zarembka, P. (Ed.). *Frontiers in Econometrics.* Academic Press. New York.

Geisser, S., and Eddy, W. F. (1979). A predictive approach to model selection. *Journal of the American Stat. Assn.*, **74**, 153–160.

Heeler, R. M., and Hustad, T. P. (1980). Problems in predicting new product growth for consumer durables. *Management Science*, **26**, 1007–1020.

Jones, K. F., and Kay, N. (1979). Pearl's Law and the growth of the Scottish National Party. *Bulletin in Applied Statistics*, **6**, 195–214.

Lawles, J. F., and Singhal, K. (1979). The efficient screening of non-normal regression models. *Biometrics*, **34**, 318.

Ljung, G. M., and Box, G. E. P. (1978). On a measure of lack of fit in time series models. *Biometrika*, **65**, 2, 297–303.

Neave, H. R. (1978). *Statistical Tables.* Allen & Unwin, London.

Pesaran, M. H. (1974). On the general problem of model selection. *Rev. Econ. Stud.*, **41** (26), 153–172.

Pregibon, D. (1980). Goodness of link tests for generalized linear models. *Appl. Statist.*, **29**, No. 1, 15–24.

Schwarzenbuck, J. K. and Gill, K. F. (1978). *System Modelling and Control.* Edward Arnold. London.

Senge, P. M. (1977). Statistical estimation of feedback models. *Simulation*, **28**, 177–184.

Tukey, J. W. (1977). *Exploratory Data Analysis.* Addison Wesley. Reading, Mass.

Wellstead, P. E. (1979). *Introduction to Physical System Modelling.* Academic Press. New York.

Wood, E., and Iturbe, I. R. (1975). A Bayesian approach to analysing uncertainty among flood frequency models. *Water Resources Rs.*, **11**, 839–843.

Part IV Estimation

Bard, Y. (1974). *Nonlinear Parameter Estimation.* Academic Press. New York.

Beck, J. W., and Arnold, K. J. (1977). *Parameter Estimation in Engineering and Science.* John Wiley. New York.

Box, G. E. P., and Tiao, G. C. (1973). *Bayesian Inference in Statistical Analysis.* Addison-Wesley. Reading, Mass. U.S.A.

Daniel, C., and Wood, F. S. (1980). *Fitting Equations to Data.* 2nd edn. John Wiley. New York.

Edwards, A. W. F. (1972). *Likelihood.* Cambridge University Press. London.

Fedra, K., Van Straten, G., and Beck, M. B. (1981). Uncertainty and arbitrariness in ecosystems modelling: a lake modelling example. *Ecological Modelling*, **13**, 87–110.
Wasan, M. T. (1970) *Parametric Estimation*. McGraw-Hill. New York.
Zellner, A. (1971). *An Introduction to Bayesian Inference in Econometrics*. John Wiley. New York.

Part V Validation

Davis, N. and Newbold, P. (1980). Forecasting with misspecified models. *Appl. Statist.*, **29**, 87–92.
Herman, C. (1967). Validation problems in games and simulations. *Behavioural Science*, **12**, 216–230.
Ignall, E. J., Kolesar, P., and Walker, W. E. (1978). Using simulation to develop and validate analytic models—some case studies. *Operations Research*, **26**, No. 2, 237–253.
Jacoby, S. L. S., and Kowalik, J. S. (1980). *Mathematical Modelling with Computers*. Prentice-Hall. Englewood Cliffs, New Jersey.
Kashyap, D., and Rao, A. R. (1976). *Dynamic Stochastic Models from Empirical Data*. Academic Press. New York.
Marriott, F. H. C. (1979). Barnard's Monte Carlo tests: how many simulations. *Appl. Statist.*, **28**, No. 1, 75–77.
Maybeck, P. S. (1979). *Stochastic Models Estimation and Control*. Vol I. Academic Press. New York.
Naylor, T. H. (1971). *Computer Simulation Experiments with Models of Economic Systems*. John Wiley. New York.
Rose, M. R., and Harmsen, P. (1978). Using sensitivity analysis to simplify models: a case study. *Simulation*, **31**, No. 1, 15–26.
Tavares, L. V. and Lopes, J. A. (1980). A strategic model for intervention on the Portuguese olive-oil market. *J. Opl. Res. Soc.*, **31**, 813–823.

Part VI Application

Aitchison, J., and Dunsmore, I. R. (1972). *Statistical Prediction Analysis*. Cambridge University Press. London.
Batty, M. (1976). *Urban Modelling: Algorithms, Calibrations, Predictions*. Cambridge University Press. London.
Bliss, C. I. (1970). *Statistics in Biology*. McGraw-Hill. New York.
Burghes, D. N., and Wood, A. D. (19). *Mathematical Models in the Social Management and Life Sciences*. Ellis Horwood. Chichester, England.
Chatfield, C. (1980). *The Analysis of Time Series, Theory and Practice*. 2nd edn. Chapman & Hall. London.
Coffman, C. V., and Fix, G. J. (Eds). (1979). *Constructive Approaches to Mathematical Modelling*. Academic Press. London.
Gilchrist, W. G. (1976). *Statistical Forecasting*. John Wiley. Chichester, England.
Greenberger, M., Crenson, M. A., and Crissey, B. C. (1976). *Models in the Policy Process*. Russel Sage Foundation, New York.
Gold, H. J. (1977). *Mathematical Modelling of Biological Systems*. John Wiley. New York.
Gottschalk, L., Lindh, G., and de Mare, L. (1977). *Stochastic Processes in Water Resources Engineering*. Water Resource Publication. Fort Collins, USA.
Guetzkow, H., Kotler, P., and Schultz, R. (1972). *Simulation in Social and Administrative Science*. Prentice-Hall, Englewood Cliffs, New Jersey.

Koutsoyiannis, A. (1977). *Theory of Econometrics.* 2nd edn. Macmillan Press. London.

Meadows, D. L., and Meadows, D. H. (Eds). (1973). *Towards Global Equilibrium: Collected Papers.* Wright-Allen Press. Cambridge, Mass.

Minshull, R. (1975). *An Introduction to Models in Geography.* Longman. Harlow, Essex, England.

Naert, P., and Leeflang, P. (1978). Building implementable marketing models. *Martinus Nijhoff.* Leiden.

Pindyck, R. S., and Rubinfield, D. L. (1976). *Econometric Models and Economic Forecasts.* McGraw-Hill. New York.

Reif, B. (1973). *Models in Urban and Regional Planning.* Leonard Hill Books. London.

Sinclair, M. A., and Dury, C. G. (1979). On mathematical modelling in ergonomics. *Applied Ergonomics,* **10,** 4, 225–234.

Schwarzacher, W. (1975). *Sedimentation Models and Quantitative Stratigraphy.* Elsevier. Amsterdam.

Zinnes, D. A. (1976). *Contemporary Research in International Relations.* Free Press, New York.

Part VII Iteration

Andrews, D. F., and Pregibon, D. (1978). Finding the outliers that matter, *J. R. Statist. Soc.* B, **40,** 85–93.

Beltrami, E. J. (1977). *Models for Public Systems Analysis.* Academic Press. New York.

Bibby, J., and Toutenberg, H. (1977). *Prediction and Improved Estimation of Linear Models.* John Wiley. Chichester. England.

Carey, D. I. & Haan, C. T. (1975). Using Parametric Models of Runoff in Improve Parameter Estimates for Stochastic Models. Water Resources Research, **11,** No. 6, 874–878.

Carbone, R., and Lougini, R. L. (1977). A feedback model for automated real estate management. *Management Science,* **24,** 241–248.

Jarboe, J. E., and Haan, C. T. (1974). Calibrating a water yield model for small ungaged watersheds. *Water Resour. Res.,* **10**(2), 256–262.

Launer, R. L., and Wilkinson, G. N. (Eds). (1979). *Robustness in Statistics.* Academic Press. New York.

Maybe, J. S., and Noel, D. V. (1980). Long-term forecasting of smooth approximations to the load duration curve. *Applied Mathematical Modelling,* **4,** No. 2, 130.

Nettheim, N. F. (1964). Fourier methods for evolving seasonals *J. Amer. Statist. Assn.,* **60,** 492–502.

Young, R. (1979). Economic causes of time series characteristics of economic variables. Nottingham University paper.

Problems

Introduction

In teaching aspects of statistical modelling the author has used two approaches to setting problems that seek to add a little realism and to raise practical as well as theoretical considerations. The first is the 'In-tray' exercise. In this the student is put in the situation where his boss, a statistical consultant, is ill and he is given the letters from his 'in-tray' to reply to. In the second the student is given a very simple practical experiment to carry out and asked to devise a suitable model for the data obtained. These approaches have been used in this short section of problems.

Experiment—whizz

Allow a toy car to run a distance x down a ramp on to level ground and measure the distance y it travels along the horizontal. Fit and validate a linear regression model of y on x.

Experiment—fizz

Obtain a tin of the type of health salts that fizz when put in water. Use a very small spoon to place n spoonfuls on a sheet of paper. Put a fixed quantity of water in a glass and empty in the salts. Time the fizzing that takes place. Repeat for $n = 1, 2, \ldots, 6$. Repeat the whole experiment for water at 5 different temperatures. Identify and fit a model relating 'Fizz Time' to quantity and temperature.

Experiment—target

Set up a target on the floor, a small box will do. Roll 8 marbles, or small balls, at it from a fixed distance, x_i, away. Count the number of hits, r_i. Repeat this for twenty distances ranging from very close to far away. Do the distances in random order to minimize problems of improvement with practice. Devise and fit a model for this situation.

Experiment—skill

Buy a simple puzzle and do it twenty times in succession, timing yourself each time. Identify, fit, and validate a suitable model.

Experiment—lift

Count the number of people getting off lifts on the ground floor of a building over a mid-morning or mid-afternoon period.
 Identify, fit, and validate a suitable model.

Experiment—lights

Count the number of vehicles passing in one direction through a set of traffic lights during 'green' phases. Assume and fit a Poisson Model. Validate your model.

Experiment—predator

There is evidence that some predators seek their prey by a random search. This can be simulated by putting at random a number of discs on a flat surface of about card table size. These can then be sought blindfold by poking a finger vertically. Remove any discs found. Count the number of 'prey' caught in one minute (or, if there are difficulties in timing, using 60 'pokes'). Repeat the experiment with varying numbers of discs on the table. Use 5, 10, 20, 30, ... Plot the numbers caught against the 'prey density'. Fit a suitable model to this data. Discuss the quality and form of the data. Discuss and if necessary carry out further experiments to find the distribution of numbers caught for some fixed density.

1. Ace transport limited

Dear Sam,

I have developed a simple little model for predicting the distance my lorries have to cover between points, I call it d, from the distance measured on a map (in a straight line) which I call x. The model is

$$d = cx$$

My problem is how to calculate c from a set of data. I used to do it by plotting and drawing a line by eye. The constant c was then the slope of the line. I have been persuaded that I should do it mathematically but different 'advisers' suggest different methods. I, at the moment, have

$$\text{Option 1} \quad \text{Use } c = \frac{\text{average } d}{\text{average } x}$$

Option 2   Use c = average slope $\left(\dfrac{d}{x}\right)$.

Option 3   —a mysterious formula $c = \dfrac{\sum dx}{\sum x^2}$

Are these all different? What should I use? Why?

Help

FRED

2.   Line Lumber Co. Ltd

Dear Sir,

We have collected the following data: l = length of standard planking obtained from a tree

d = diameter of tree at breast

l	d
960	35
564	27
261	21
1227	36
284	24
89	17
705	30
168	19
431	24
104	18
1184	38
907	32
117	16
1048	36
284	23
1322	40
449	27
615	31
1479	40
143	18

Can you suggest a suitable way of predicting l from d?
Yours sincerely,
TREVOR JONES

3.   **J. G. Limited**

SERVICE DEPARTMENT

Dear Sam,

We are thinking about the size of our repair team. I would like to do some simulations but I need to model the distributions involved. I have collected the following data:

Time between calls for repairs (in hours)

time	frequency
0–5	86
5–10	41
10–15	34
15–20	16
20–25	8
25–30	8
30–35	1
35–40	3
40–45	2
45–50	1
50 +	0

Time to carry out repair (ignoring trivial ones taking less than one hour)

time	frequency
1–2	1
2–3	0
3–4	5
4–5	7
5–6	16
6–7	31
7–8	43
8–9	35
9–10	31
10–11	20
11–12	10
12–13	3
13–14	1
14 +	0

Can you please suggest suitable models and constants to use in the simulation.

Best wishes,

JIM

4.   Biology department

Dear Sam,

I am still collecting moths, in fact counting them in rather large quantities. I have a number of sample areas round the country and, with the help of some students, we have done some very thorough counts on the numbers in the trial areas at the same time in two successive years. As you will see from the data some of the areas have the moths we are looking at in plague proportions. We suspect that the weather may influence reproduction so I have also included a rough temperature measure of deviations from the norm. The data is

Number in first year	Number in second year	Temperature factor
63	69	-3.8
79	400	3.0
398	3162	3.5
501	1010	0.5
1000	1270	-0.5
1259	316	-3.5
1585	501	-2.0
1585	6309	0.5
2512	12590	2.5
3162	3170	1.0
3981	2510	-1.0
5012	3172	0.5
6309	6328	-0.5
12590	6290	0.0
15850	12640	-1.5
15930	6590	-2.5
31620	19950	1.0
63090	7943	-3.5

Ignoring the temperature factor initially what do you think is the relation between the numbers in successive years. Secondly how does the temperature influence things? What would happen if the temperature was above or below average for a number of successive years?

Yours

ARTHUR JONES

5.   **Memorandum**

To: Sam Smith

From: Jim Able

You may remember that I talked to you about the new glass vessels when they were in the development stage. We were concerned about the presence of bubbles and you suggested the Poisson model as appropriate. At that stage we decided that if the mean number of bubbles per vessel was less than 0.5 we would accept the batch. It all seemed OK at the time. The quality people took samples from each batch and worked out the average. My problem is that we have become too clever. We now have a light device that scans each vessel on the production line and indicates the presence or absence of bubbles. The trouble is that it cannot count the bubbles, so how do I estimate the mean?

6.   **St James School**

Dear Sam,

More school problems I'm afraid. I have started using multiple choice exams quite a lot, but I always worry about guessing. For example suppose I set 24 questions each with 4 possible answers then by sheer chance one could get 6 answers right, so if a pupil gets 18 I am inclined to call it 12. What I would like is some proper argument. If I assume he knows n and guesses m in an $n+m$ question paper and he gets r right—how can I estimate n? For generality I have assumed there were k possible answers given for each question. I didn't get very far with my own efforts.
Any suggestions.

Yours

MIKE

7.   **S. M. Manufacturing**

Marketing Division

Dear Sam,

We have recently done a study that was intended to find out what proportion, P, of our 'Target' firms might actually use our new equipment. We recorded the time from the launch of the new product to first purchase and recorded these as $t_1, t_2, \ldots$ for the 20 firms out of the 80 Targets. The average time was 8 months. We only kept the study going for 24 months. The problem is those who didn't buy in the 24 months. Had we gone on longer they might have bought, given time, so I cannot estimate P as 20/80. I have looked through the technical literature and found a formula for the probability (?) for such a situation—at least I think it is. It looks terrible—

$$(P\alpha)^m e^{-m\bar{a}\bar{t}} (1 - P + Pe^{-\alpha T})^{n-m}.$$

I think $m = 20$, $n = 80$, $T = 24$ months, $\bar{t} = 8$ months. I don't know what α means. Can you explain this in words of one syllable? Even if it is right how can I find P from it?

Yours

JIM

8. **St John's School**

Dear Sir,

Our geology students have been taking measurements on the thickness of a local bed of sedimentary rock. One theory is that it is actually composed of a small number of sub-beds, produced during exceptional but similar conditions. We lack the equipment to detect these. It was suggested to me that the distribution of thickness might actually reflect the number of sub-beds. The data we have, from well separated measurements, is, in centimetres,

6.1, 5.4, 6.8, 7.2, 5.9, 6.3, 4.9, 7.5, 5.8, 6.7.

Can you please suggest how many sub-beds there might be.

Yours sincerely,

F. TAYLOR
Head—Geography Department

[Note (a) the sum of α exponentials is a two-parameter Gamma distribution—Tables 3.3 and 3.4
(b) For integer α $\Gamma(\alpha) = (\alpha - 1)\Gamma(\alpha - 1) = (\alpha - 1)!$
The differentiation of $\Gamma(\alpha)$ can be avoided by using numerical exploration.]

326

9.   S. M. Manufacturing

Quality Control Division

Dear Sam,

We have an automated production line that produces metal strips ten metres long. Flaws may occur in the strip which are detected by an x-ray scanner under which the strip passes. If no flaws are detected the strip continues on down the production line. If a flaw is detected the strip is marked automatically. If there is only one flaw the strip is stacked for re-use. If a second flaw is detected the strip is immediately removed and scrapped and no attempt is made to seek for further flaws.

For example in a run of 100 strips 91 passed, 8 were stacked and 1 was scrapped. Could you suggest how this situation could be modelled and how I might use the model to predict the percentage scrap that we are likely to have long term.

Best wishes,

Fred

10.   S. M. Manufacturing

Test Division

Memo: Sam

We have been carrying out long-term failure tests of a component. We are fairly sure from past experience that the mean time to failure is 2.60 units (unit = 10,000 hours). To assign reliabilities we need to know the distribution. We have always assumed an exponential but a modified Pareto distribution has been suggested.

$$\text{i.e. } f(t) = \alpha(t+1)^{\alpha-1} \qquad t > 0, \quad \alpha > 2$$

The only data I have is

0.56 1.25 1.87 7.66 6.02
0.24 0.29 4.60 3.23 1.19

Any suggestions?

Roger

Hints and comments on experiments and letters

Experiments

For most of the experiments it is possible to obtain a conceptually identified model. This should be derived before carrying out the experiment as it may suggest ways of improving the experiment. For example, in Experiment—Target use can be made of the geometry of the Experiment together with the reasonable assumption that the angle of throw, with the centre of target, is Normal with a standard deviation that is larger than the angle subtended by the target. This leads to the approximation that for larger distances (d) the probability of hitting the target depends on $1/d$. The method of maximum likelihood enables estimates to be obtained and the χ^2 goodness of fit test provides a means of validation.

Letters

1. Ace Transport Limited: Modify the model to include an error term:

$$d_i = cx_i + e_i, \quad E(e) = 0 \quad V(e) = \sigma^2.$$

Substitute this in the three estimator formulae, for example

$$c_1 = \frac{\Sigma d_i}{\Sigma x_i} = \frac{c\,\Sigma x_i + \Sigma e_i}{\Sigma x_i} = c + \frac{\Sigma e_i}{\Sigma x_i}.$$

Hence

$$E(c_1) = c \quad V(c_1) = \frac{1}{(\Sigma x)^2}\, V(\Sigma e_i) = \frac{n}{(\Sigma x)^2}\sigma^2.$$

All three estimators are unbiased. The three variances can be compared via mathematics or some numerical trials. The Least Squares estimator c_3 has the smallest variance.

2. Line Lumber Co. Ltd: A conceptual model can be found by assuming the cut tree is approximately cylindrical of length k and radius r. If it is further assumed that the shape of the type of tree is independent of size then the length of the useful log will be proportional to the radius,

327

$k \propto r$. Hence the volume and thus the length of planking obtainable will depend on $\pi r^2 . k$, i.e. on r^3. The model suggested is thus length $\propto r^3$, which in turn suggests plotting ln (length) against r in the empirical identification.

3. J. G. Limited
Time between calls—exponential.
Repair times—normal.

4. Biology Department
N_t in year t. Initial model

$$\ln N_{2i} = \alpha + \beta \ln N_{1i} + e_i.$$

Fit line and plot residuals against temperature factor.

5. Memorandum
From Poisson Pr (no bubbles) $= e^{-\mu}$
$\quad\quad\quad Pr$ (one or more bubbles) $= 1 - e^{-\mu} = \theta$ say.
From n bottles, assuming independence of bubble numbers between bottles, then the number of bottles showing bubbles, r, will have a Binomial distribution Bin (n, θ). So $\hat\theta = r/n$ is the maximum likelihood estimator of θ and that for μ will be given by $1 - e^{-\hat\mu} = \hat\theta$. Thus $\hat\mu = \ln(1 - r/n)$. A practical consideration here is the possible use of a tracking signal to monitor a process that now has 100% 'sampling'.

6. St James School: In $N = n + m$ questions r are correct, n because of knowledge, s by guessing, so $r = n + s$. The probability of the data $= Pr(s)$, treating n as the fixed but unknown parameter.

The distribution of s is Binomial Bin $(m, \frac{1}{k})$.

$$\text{The likelihood } \mathscr{L}(n) = \binom{m}{s}\left(\frac{1}{k}\right)^s\left(1 - \frac{1}{k}\right)^{m-s}$$
$$= \binom{N-n}{r-n}(1-k)^{N-r}/k^{N-n}.$$

This can be evaluated numerically for different n to find the maximum likelihood estimator. The method of moments gives a simpler estimator since

$$E(r) = n + E(s) = n + \frac{m}{k} = n + \frac{N-n}{k}.$$

Equating this with the observed r gives

$$\hat n = \left(r - \frac{N}{k}\right)\left(\frac{k}{k-1}\right).$$

7. S. M. Manufacturing: The 'probability of a time' will actually be the probability density of t given that the 'Target' firm buys multiplied by the probability P that the firm does buy. The simplest model for time to first purchase is an exponential model. Thus the contribution to the likelihood from the 20 times is

$$(\lambda e^{-\lambda t_1} \cdot P)(\lambda e^{-\lambda t_2} \cdot P) \ldots (\lambda e^{-\lambda t_{20}} \cdot P).$$

For the 60 firms that did not buy there is a probability P that they are in the class of potential buyers but would buy after 24 months, for which the probability is $e^{-\lambda \, 24}$. Alternatively there is a probability $1 - P$ that they would never buy. Thus the probability for a firm not to have bought is

$$Pe^{-24\lambda} + 1 - P.$$

Thus the likelihood is

$$\mathscr{L}(\lambda) = (\lambda^{20} e^{-\lambda 20\bar{t}} P^{20})(Pe^{-24\lambda} + 1 - P)^{60}.$$

This can be solved numerically using the methods discussed in Chapter 13.

8. St John's School: Use the note and assume the sub-bed thickness x_i to be exponential with constant mean μ. For α sub-beds the total thickness will be

$$y = \sum_{1}^{\alpha} x_i.$$

Hence

$$E(Y) = \alpha E(X) = \alpha\mu, \quad V(Y) = \alpha V(X) = \alpha\mu^2,$$

assuming independence of the x_i. From this $E(Y)^2/V(Y) = \alpha$. Thus a method of moments estimator of α is $\bar{y}^2/s_y^2$. The method of maximum likelihood uses the likelihood for the two-parameter Gamma ($\mu = 1/\lambda$), differentiates with respect to α and equates to zero. This gives $\hat{\lambda} = n\hat{\alpha}/\Sigma x$ from one equation. The other equation is highly complex. The procedure therefore is to substitute for $\hat{\lambda}$ in the likelihood and numerically evaluate the likelihood for integer values of $\hat{\alpha}$ around that given by the method of moments. The α for maximum is thus found. Note the very strong assumptions that are required for this model.

9. S. M. Manufacturing: Assume Poisson process of flaws per 10-metre strip.

$$Pr(0 \text{ flaws}) = e^{-\mu}, \quad Pr(1 \text{ flaw}) = \mu e^{-\mu}$$
$$Pr(\text{more than one flaw}) = 1 - e^{-\mu}(1 + \mu).$$

Hence likelihood on sorting out for the given data

$$\mathscr{L}(\mu) = e^{-100\mu} \mu^8 [e^{\mu} - (1 + \mu)].$$

A first estimate of μ comes from Pr (0 flaws) $= e^{-\mu}$ which corresponds to 0.91 of the data. The iterative solution of subsection 13.4.1 can then be applied.

10. S. M. Manufacturing: This can be tackled informally using Probability Plots that can be constructed for both distributions or by looking at initial likelihood. A Bayes decision or dispersion test could also be looked at.

Appendix 1

Probability and probability laws

Most people have some intuitive ideas of probability. We assign low probability to events that we do not expect to happen and vice versa. Thus probability is a property assigned to events in an intuitive modelling process. There are many ways of making this intuition more formal. We consider the three most common.

(a) If we take a purely conceptual approach we can make progress if we limit ourselves to events involving an inherent symmetry, throwing dice, choosing cards etc. Thus if we have some situation with n symmetric outcomes, a symmetric n-sided die is thrown, of which r are identified as giving an event E, for example $E = (\text{score} \leqslant r)$, *then we define*

$$Pr(E) = \frac{r}{n},$$

as a conceptual definition of this property called probability.

(b) If we take an empirical approach we can make a number of trials, n, and count the number of occasions, r, on which the event E occurs. The relative frequency, $Rf(E) = r/n$ of E corresponds empirically to our idea of probability. The problem here is that different sets of n observations will give different values of r. The way out of this is to take larger and larger values of n and see if $Rf(E)$ tends to a stable value, $Pr(E)$. If we can measure or assume such a limit we call $Pr(E)$ the probability of E.

(c) We shall often need to refer to the chances of an event E occurring at a trial that is unique. Indeed even 'the next trial' in the sequence of (b) is a unique event and we can only use the ideas of (b) subject to assumptions about the constancy of the environment and the prototype. We thus bring into the situation our assessments and beliefs. An approach that emphasizes this aspect simply states that for a trial we can express our

331

subjective degree of belief of E occurring as a number $Pr(E)$ that we call the probability. This assessment may well be based on symmetry arguments, relative frequency evidence from similar situations, etc., but finally the number is a measure of our belief about the possible occurrence of E.

Whichever of these approaches we adopt the quantity $Pr(E)$ will need to possess certain reasonable properties:

(a) If an event E is an impossibility, $E = I$, then

$$Pr(I) = 0,$$

conversely if it is a certainty, $E = C$, then

$$Pr(C) = 1.$$

In general $0 \leqslant Pr(E) \leqslant 1$.
(b) Often events are made up of a number of component events.
   Ex (a) Consider the number of successes, R, in an experiment involving six attempts at a task.
       The event: (R is even) is composed of the component events ($R = 2$), ($R = 4$), ($R = 6$)
       so (R = even) $\equiv$ ($R = 2$ or $R = 4$ or $R = 6$).
   Ex (b) The event: (successive experiments both give $R = 2$) is composed of two outcomes ($R_1 = 2$) and ($R_2 = 2$).
   Ex (c) The event: (r is both even and less than 3) contains two obvious components, (R even) and ($R < 3$).

In situations such as these it is helpful to show pictorially the possible values of the random variables in a *Sample Space*. Figure A.1 illustrates. In the sample space the nature of particular events can be represented by a subspace, illustrated in Figure A.1 by (a) (b) and (c) for the three examples respectively.

   The question of obvious interest is how the probability of an event is related to the probabilities of the component events. The answers to this question depend on the interrelationships between the components. Suppose first of all that there is no relationship between the component events, that they are *Independent*. In this case, if we have two independent events A and B then, the probability of them both occurring is

$$Pr\,(A \text{ and } B) = Pr(A) \times Pr(B)$$

This formula is effectively the definition of the term independence and is called the *Multiplication Law*. In Ex (b) with two independent experiments

$$Pr(R_1 = 2 \text{ and } R_2 = 2) = Pr(R_1 = 2) \times Pr(R_2 = 2).$$

If the events are related then this law needs modification. To introduce one

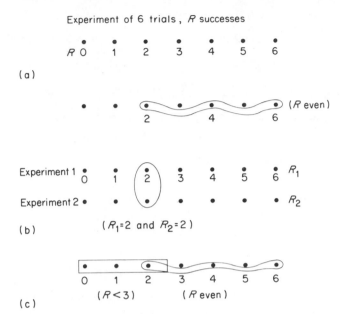

Figure A.1 Sample spaces

modification we need the idea of conditional probability. The *Conditional Probability*, $Pr(A|B)$, is the probability of A occurring given that B has already occurred. This, and the opposite quantity $Pr(B|A)$, indicates the relations between A and B. If A and B are independent, clearly $Pr(A|B) = Pr(A)$ and vice versa. In the general case

$$Pr(A \text{ and } B) = Pr(A|B)Pr(B)$$
$$= Pr(B|A)Pr(A).$$

Thus the probability of both A and B occurring equals the probability of B first occurring multiplied by the probability of A occurring given that B has occurred, or vice versa.

One special case is that where both A and B cannot occur together

$$Pr(A|B) = Pr(B|A) = 0.$$

Here A and B are said to be *Mutually Exclusive* events. We know that $Pr(A \text{ and } B) = 0$ but another probability of interest is $Pr(A \text{ or } B)$, for example $Pr(R = 2 \text{ or } R = 4 \text{ or } R = 6)$ in Ex (a). A consideration of the sample space suggests the following rule, called the *Addition Law*,

$$Pr(A \text{ or } B) = Pr(A) + Pr(B).$$

In the example

$$Pr(R = 2 \text{ or } R = 4 \text{ or } R = 6) = Pr(R = 2) + Pr(R = 4) + Pr(R = 6).$$

Note that, taking the whole sample space,

$$Pr(R = 0 \text{ or } R = 1 \text{ or } R = 2 \text{ or } R = 3 \text{ or } R = 4 \text{ or } R = 5 \text{ or } R = 6) = Pr(C)$$

so

$$\sum_{\text{all } t} Pr(R = r) = 1.$$

This is sometimes called the *Total Probability Rule*. If the events A and B are not mutually exclusive the Addition Law becomes

$$Pr(A \text{ or } B) = Pr(A) + Pr(B) - Pr(A \text{ and } B).$$

The logic of this can be seen by looking at the sample space for Ex (c). In this there is an overlap for the sample space for (R is even) and ($R < 3$). Thus, intuitively, adding the two probabilities double counts this overlap, which is the event [(R is even) and ($R < 3$)]. This event is in fact ($R = 2$) so the formula gives

$$Pr[(R \text{ even}) \text{ or } (R < 3)] = Pr(R \text{ even}) + Pr(R < 3) - Pr(R = 2).$$

One final formula is worth recording: this relates to both dependent and mutually exclusive events. Suppose after completing the six trials a further task is carried out. The probability of success S at this task is assumed to depend on the number of successes in the six trials, now

$$Pr(S) = \sum_{\text{all } r} Pr(S \text{ and } r)$$

since S must occur with one of the r values. But

$$Pr(S \text{ and } r) = Pr(S|r) \, Pr(r)$$

so

$$Pr(S) = \sum_{\text{all } r} Pr(S|r) \, Pr(r).$$

The general formula of this form is called the *Chain Rule*.

Appendix 2

Approximations

In severa places in this book we have replaced a curve by an approximating straight line at some point z_0, for example in Figure 13.2. If the function of the curve is $g(z)$ then this approximation can be written as

$$g(z) \simeq g(z_0) + (z - z_0). \text{ slope of } g(z) \text{ at } z_0$$

i.e.

$$g(z) \simeq g(z_0) + (z - z_0) \left(\frac{dg}{dz}\right)_{z_0}.$$

We have used this in solving equations in Chapter 13. Another important use is in finding approximations to the expectations of functions. Thus, if z is a random variable with expectation μ, and if the random variation of z is such that $g(z)$ is approximately linear in the region where observations are most probable, then we can write

$$g(z) \simeq g(\mu) + (z - \mu) \left(\frac{dg}{dz}\right)_{\mu}$$

so

$$E(g(z)) \simeq g(\mu)$$

and

$$V(g(z)) \simeq V(z) \cdot \left(\frac{dg}{dz}\right)_{\mu}^2.$$

For a curve that is not approximately linear we extend the series of terms and have

$$g(z) \simeq g(z_0) + (z - z_0) \left(\frac{dg}{dz}\right)_{z_0} + \frac{(z - z_0)^2}{2!} \left(\frac{d^2g}{dz^2}\right)_{z_0} + \frac{(z - z_0)}{3!} \left(\frac{d^3g}{dz^3}\right)_{z_0} + \dots$$

This series is called the Taylor series. The first three terms were used in an approximation in subsection 21.1.2.

Index